Taxonomy          Phylogenetic Tree
                    (Tree of Life)

| Domain |
| Kingdom |
| Phylum |    (Phyla - plural)
| Class |
| Order |
| Family |
| Genus |    (Genera - plural)
| Species |

|         | Whales | Seals | Eared Seals | Walruses | Manatees & Dugongs |
|---------|--------|-------|-------------|----------|--------------------|
| Domain  |        |       |             |          |                    |
| Kingdom | Animalia |     |             |          |                    |
| Phylum  | Chordata |     |             |          |                    |
| Class   | Mammalia |     |             |          |                    |
| Order   | Cetacea | Pinnipedia | | | Sirenia |
| Family  | | Phocidae, Otariidae, Odobenidae | | | |

# Exploring Creation

## with

# Zoology 2

## Swimming Creatures of the Fifth Day

by Jeannie K. Fulbright

**Exploring Creation with Zoology 2: Swimming Creatures of the Fifth Day**

*Published by*
*Apologia Educational Ministries, Inc.*
*1106 Meridian Plaza, Suite 220/340*
*Anderson, IN 46016*
www.apologia.com

Manufactured in the United States of America
Ninth Printing
February 2016

ISBN: 978-1-932012-73-6

**Printed by R.R. Donnelley**

*All Biblical quotations are from the New American Standard Bible (NASB)*

*Cover photos © Dennis Sabo, © Peter Evans, © Bartlomiej Kwieciszewski, © Stephanie Phillips,*
*© Peter Heiss, © Steffen Foerster, © Daniel Gustavsson, © Julie Wax, © Getty Images*

*Cover design by Kim Williams*

# Apologia's Young Explorer Series
# INSTRUCTIONAL SUPPORT

Apologia's elementary science materials launch young minds on an educational journey to explore God's signature in all of creation. Our award-winning curriculum cultivates a love of learning, nurtures a spirit of exploration, and turns textbook lessons into real-life adventures.

## TEXTBOOK

**Apologia Textbooks** are written directly to the student in a highly readable conversational tone. Periodically asking students to stop and retell what they have just heard or read, our elementary science courses engage students as active learners while growing their ability to communicate clearly and effectively. With plenty of hands-on activities, the Young Explorer Series allows young scientists to actively participate in the scientific method.

## NOTEBOOKING JOURNAL

Spiral-bound **Apologia Notebooking Journals** contain lesson plans, review questions, full-color mini-books, puzzles, and much more to keep students actively engaged in learning while keeping them organized.

## JUNIOR NOTEBOOKING JOURNAL

Designed for younger students and those who struggle with writing, **Apologia Junior Notebooking Journals** cover everything in the regular notebooking journals, but at a more basic level and with primary writing lines. With simpler vocabulary pages, and additional coloring pages, junior notebooking journals make science enjoyable for even your youngest student.

## AUDIO BOOKS

Some students learn best when they can see and hear what they are studying. Having the full audio text of your course is great for listening while reading along in the book or riding in the car! **Apologia Audio Books** contain the complete text of the book read aloud to your student.

## FIELD TRIP JOURNAL

**Apologia's Field Trip Journal** is a fun and exciting way to record those moments when textbook lessons turn into real-life adventures. You and your students can successfully prepare for field trips, map the places you visit, and document entire field trips from planning stage to treasured memory.

At Apologia, we believe in homeschooling. We are here to support your endeavors and to help you and your student thrive! Find out more at apologia.com.

# Scientific Speculation Sheet

**Name** _____  **Date** _____

**Experiment Title** _____

Materials Used:

Procedure: (What you will do or what you did)

Hypothesis: (What you think will happen and why)

Results: (What actually happened)

Conclusion: (What you learned)

# Introduction

Welcome to the wonderful world of underwater creatures! *Exploring Creation with Zoology 2: Swimming Creatures of the Fifth Day* will take you on a journey through the oceans and streams of the world to discover many of the charming, exotic, fascinating, and fanciful creatures God created on the fifth day.

You will find this to be an easy-to-use science curriculum for your entire family. The text is written directly to the student, making it very appealing for elementary students. The material is presented in a conversational, engaging style that will make science enchanting and memorable for your students, creating an environment in which learning is a joy.

### Zoology 2 Prerequisite

This course on aquatic animals follows the first book in the series, *Exploring Creation with Zoology 1: Flying Creatures of the Fifth Day*. Although it is recommended that you begin your study of Zoology with Zoology 1, it is not required. Because foundational concepts, such as animal classification, nomenclature, instincts, and endangered species are explained in the first book, we have made that first lesson available on our website. Attainment of this knowledge is assumed in Zoology 2. We do recommend that you cover this material before you start this text.

### Lesson Increments

There are thirteen lessons in this text. Each lesson should be broken up into manageable time slots, depending on your child's or children's age and attention span. This will vary from family to family.

Most lessons can be divided into two-week segments. You can do the reading and the notebook assignments during the first week, and you can do the experiments and the data recording during the second week. If you do science two or three days per week, you might read four to six pages a day to finish a lesson and begin the experiment. This will give you thirty-two weeks for the entire book. Older students can work through the book more quickly if they wish. Sixth graders are encouraged to read the book on their own in preparation for the independence with which they will do science the following year if they use *Exploring Creation with General Science.*

When you have finished reading a lesson, the student will orally relate the information they learned (narration) and complete a creative assignment associated with the lesson (notebooking). Narrations and notebooking replace the traditional and less-effective method of filling in blanks in a workbook. I believe this is a superior method of facilitating retention and providing documentation of your child's education. For a more detailed discussion of how to use this course, please see the step-by-step guide at the end of this introduction.

## Narrations

Older elementary students can do the entire book and most experiments on their own, while younger students will enjoy an older sibling or parent reading it to them. Each lesson begins with a reading of the text. Throughout the reading, the students will be asked to retell or narrate the information they have just studied. This helps them assimilate the information with their minds. The act of verbalizing in their own words what they have learned propels them forward in their ability to effectively and clearly communicate with others that which they know. It also serves to lock the information into their minds.

## Notebooks

In addition to the written narrative, notebook activities promote further experience with the material. Each notebook activity requires the student to utilize the material he learned in a creative way that will further enhance his retention. The notebook activities generally occur at the end of a lesson. You can create your own notebook with paper and binders, or purchase the *Zoology 2 Notebooking Journal*, which has all the pages needed to complete each assignment. The *Zoology 2 Notebooking Journal* also contains a schedule, additional vocabulary work, Scripture copywork, colorful lapbook-style miniature books, extra projects, experiments, and DVD and reading suggestions.

## Project Ocean Box

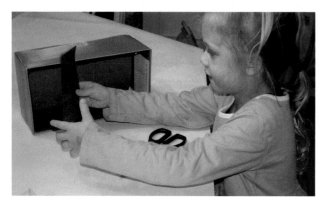

Each student or family is encouraged to create clay or paper models of the animals they have studied. The first lesson begins with the students using a box to create an "aquarium" in which they will display their animal creations. I call this aquarium an "ocean box." After completing each lesson, the students will add the animals about which they learned to the ocean box. At the end of the course, they will have a tangible model of all that they learned in science throughout the year. The materials your student or family uses to create the animals for display will vary according to the preferences of your family. You can use clay, images cut out from magazines, or even plastic animal replicas, which are often available at dollar stores. Obviously, if your family wishes not to participate in this activity, you may skip this section of each lesson.

## Experiments

Every Zoology 2 lesson ends with an experiment. These experiments will help your children develop the skills needed to conduct valid and scientifically accurate experiments. It is recommended that your students complete at least a few, if not all, of these experiments so as to become familiar with the scientific method. This will further their understanding of how to perform experiments and what makes and does not make a good experiment.

The projects and experiments in this book use mostly common, household items. As a result, they are fairly inexpensive, but you will have to hunt down everything that you need. To aid you in this is a list, beginning on page xi, of the materials you will need for the experiments and projects in each lesson.

## The Immersion Approach
### Is it Okay to Spend a Year on Just a Part of Zoology?

Many educators promote the spiral or survey approach to education, wherein a child is exposed over and over again to minute amounts of a variety of science topics. The theory goes that we just want to "expose" the child to science at this age, each year giving a bit more information than was given the year before. This method has been largely unsuccessful in public and private schools, as National Center for Education Statistics (NCES) data indicate that eighth graders are consistently less than 50% proficient in science.

This method assumes the young child is unable to understand profound scientific truths. Presenting a child with scant and insufficient science fails to develop a love for the subject. If the learning is skimpy, the subject seems monotonous. The child is simply scratching the surface of the amazing and fascinating information available in science. Sadly, students taught in this way are led to believe they "know all about" that subject, when in reality the subject is much richer than they were allowed to explore.

The National Science Foundation (NSF) says the problem with our methods of teaching science is that science curriculum is "a mile wide and an inch deep." There is too much information covered, with no opportunity for true substance to be taught. That is why I recommend that kids, even young children, are given an in-depth, above their perceived grade level exploration into each science topic. You, the educator, have the opportunity to abandon methods that don't work so that your students can learn in the ways that have been proven effective. The immersion approach is the way everyone, even young kids, learn best. That is why we major in one field in college and take many classes in that field alone. If you immerse your child in one field of science for an entire year, he will develop a love for that subject and a love for learning in general.

Additionally, a child that has focused on one subject throughout an entire year is being challenged mentally in ways that will develop his ability to think critically and retain complex information. This will actually benefit the child and give him an advantage on achievement tests. He will be able to make more intelligent inferences about the right answer on science questions, as God has created an orderly world that works very similarly throughout all matters of science. A child who has not been given the deeper, more profound information will not understand how the scientific world operates.

### Course Website

If your child would like to learn more about the animals discussed in this course, there is a course website that allows the student to dig even deeper into these aspects of zoology. To go to the course website, simply type the following address into your web browser:

http://www.apologia.com/bookextras

You will see a box on the page. Type the following password into the box:

Godmadethemswim

Make sure you capitalize the first letter, and make sure there are no spaces in the password. When you hit "enter," you will be taken to the course website.

# How to Use This Book
# A Step-by-Step Guide

1.  If you have not purchased a supplies kit, you will want to scan the materials list following this guide to see what you need for the lesson you are going to do.

2.  Begin by reading the lesson to the students (older students may read the lesson themselves). There will be places during the lesson where the students are asked to "tell back" or narrate what they have learned up to that point. These are not written narrations; they are impromptu oral presentations.

3.  Occasionally there will be a "Try This!" activity wherein the learners are encouraged to get a few supplies and try a little project or experiment to demonstrate a point made in that section of the book. Ideally, the project should be done right then. However, don't be discouraged if you do not have the materials. You can always go back and do the project later.

4.  You will continue reading until you feel a natural break is at hand. Each family will differ in the amount of reading done in each session. Some families become extremely engrossed and will want to read an entire lesson. Most families will read a quarter to half the lesson. There are many places within each lesson that are natural stopping points. You decide when to stop reading. The book is designed to give a lot of flexibility with this, so that you can complete the book in a year in a way that works for your family.

5.  When you end for the day, ask your children to orally tell you what they have learned. They do not need to write anything down until they reach the end of the lesson.

6.  When you reach the end of a lesson, you will come to a "What Do You Remember?" section. This is a series of specific questions to ask your children in order to prompt their memories about the lesson. Don't expect young children to remember most of these. Don't expect older children to remember all of them. However, this is a great time to enter into discussion about what they have learned. These are also oral, not written.

7.  After your children tell you what they remember, it's time for the notebooking activity. In this activity, each child will be asked to record in writing all that she wants to remember about the lesson. I would not force her to record every detail of the "What Do You Remember?" section. Also, do not have her write down what you want her to remember. Allow her to decide what she thought was interesting and important. Let her decide what she wants to remember. For non-writers or slow writers, you can type out or write out what they tell you. If your child is struggling to recount her learning, you can encourage her with questions. Make this an enjoyable experience

without a lot of correction and nit-picking. Eventually, your child will be able to accurately and systematically recount what she learned. Many children graduate from high school never learning this skill.

8.   Occasionally, the notebooking activity will also include some sort of work beyond just recording the information they found interesting or want to remember. They might be asked to diagram something or produce a creative work associated with the subject.

9.   Many times, older students are encouraged to do further work associated with the lesson. You, the parent, will decide if your child should complete this additional assignment.

10.  The students will then add the animal or animals studied to their ocean box.

11.  The last thing students should do is the experiment for the lesson.

Let heaven and earth praise Him, the seas and everything that moves in them.
- Psalm 69:34

# Items Needed To Complete Each Lesson

Every child will need his own notebook, blank paper, lined paper, and colored pencils.

## Lesson 1

- Another person
- Two hair dryers or personal-sized electric fans
- Glitter or pepper
- A long casserole dish
- A clear plastic 2-liter bottle with a lid
- A small (the smaller the better) balloon
- A box (A large box is best, but even a shoe box will work.)
- Enough blue paper to line the inside and outside of the box
- Glue
- A large clear bowl or container (It could even be the bottom half of a plastic soda bottle.)
- A paper or Styrofoam cup
- A nail or pen to puncture a hole in the cup
- Blue and yellow food coloring
- A spoon with which to stir
- Hot water and ice-cold water

## Lesson 2

- Two pieces of cardstock or poster paper (8½" x 11" works best)
- Tape
- Two cups
- A tablespoon
- Salt
- A freezer
- Someone to help you
- Two paper (or Styrofoam) cups
- A sharp object that can make a tiny hole in the bottom of each cup (like the tip of a straightened paper clip)
- A sharpened pencil
- 30 feet of 100% cotton yarn (Synthetics tend to stretch too much.)

## Lesson 3

- A big bowl of ice water
- A large container of petroleum jelly or shortening
- Latex or dish gloves (You'll need four for every child doing the experiment.)
- A kitchen timer clock (or stopwatch)
- Someone to help you

## Lesson 4

- Chalk
- Tape measure
- A balloon
- A box
- A teaspoon
- Vanilla
- Two tadpoles (Check the course website to find where you can order them.)
- Two covered tanks (or large glass bowls with covers) that are the same size
- A lamp with a 50- or 60-watt bulb

## Lesson 5

- Chalk
- Tape measure
- A seashell
- A Cheerio (or similar piece of breakfast cereal)
- Two glasses
- A container in which to mix water and dirt
- Water
- Dirt
- Three small plastic disposable containers
- A cup of sand
- A cup of clay (Real clay is preferable, but modeling clay is okay.)
- A cup of soil from your yard (Collect your soil at least 12 inches below the surface.)
- A detailed shell of some sort (It should have ridges or other specific features. If you can't get a shell, you can use a toy that has detail in its shape.)
- A craft stick or plastic spoon for stirring
- Two cups plaster of paris
- A cup of water
- A disposable container for plaster
- A permanent marker

## Lesson 6

♦ A fish in a small bowl or a glass (A goldfish is best, because it can handle large changes in water temperature. Do not use a tropical fish, as it will not be able to handle large temperature changes.)

♦ A lamp with a bendable neck that can face down into your fish bowl (The light bulb should be 100-watt.)

♦ A large bowl of ice water into which you can place the fish bowl or glass

♦ A thermometer (It should read temperatures above and below room temperature.)

## Lesson 7

♦ A ruler

♦ A large battery (see picture on page 120)

♦ Three pieces of insulated electrical wire

♦ Two metal nails (They must conduct electricity.)

♦ Electrical tape

♦ A glass of distilled water (You can buy this at any large supermarket. Make sure it is *distilled* water, not mineral water.)

♦ A small light-bulb holder with a light bulb (These can be found in hobby stores. See the picture on page 120.)

♦ Salt

## Lesson 8

♦ Markers

♦ A blank file folder

♦ A set of index cards

♦ Dice

♦ Animal game pieces (You can make these or use small plastic animals.)

♦ Sea-Monkeys or triops (See the course website for where they can be purchased.)

## Lesson 9

♦ Several bottles of different sizes

♦ A least one shoe box

♦ Scissors

♦ Glue

♦ Some cardboard (You can cut up cereal boxes or other thin cardboard boxes.)

♦ Cotton batting or a sheet of foam

## Lesson 10

♦ A small eyedropper

♦ One empty, clear water bottle or soda pop bottle with a lid

♦ A glass of water

♦ Optional: a google eye

♦ Optional: 1 chenille stick (pipe cleaner) or yarn

♦ Optional: double-sided tape

## Lesson 11

♦ Five 6-inch segments of cotton string

♦ Three cups of water

♦ A lot of salt

♦ Food coloring (any color)

♦ A wide-mouthed glass jar (A glass measuring cup will do.)

♦ A pan for boiling water

♦ A wooden spoon for stirring

## Lesson 12

♦ A glass baking dish

♦ Hot water

♦ A fruit ice pop (For best results, use one made from real fruit, like an Edy's frozen fruit bar.)

## Lesson 13

♦ Jar

♦ A clipping from a vine or a rose

♦ Pond water

♦ Two water bottles

♦ Black paint

♦ One cup of water mixed with one tablespoon of salt

♦ Plastic tubing

♦ Clay

♦ A nice sunny window

♦ A stack of books

Exploring Creation with Zoology 2: Swimming Creatures of the Fifth Day
Table of Contents

# Lesson 1
# Aquatic Animals

From the streams that begin in the mountains, through the lakes and rivers that lead to the oceans, and far down into the depths of the sea, the Lord God filled up the waters with creatures great and small. With a word, the enormous whales sprang into being. At His command, billions of plankton leapt to life. In one moment, full-grown sea turtles, sharks, sponges, dolphins, squids, and octopuses joined them in the sea. Strongly swimming fish headed up the

This interesting creature is a nudibranch. Like all swimming creatures, it was made by God on the fifth day of creation.

streams with the creeping crayfish and the sluggardly snail. Indeed, the fifth day of the earth's existence was crammed with excitement. We will explore the wonders of this day, focusing on creatures that swim in the water, whether they are in the pond down the street or in the ocean deep. So, put on your scuba gear and let's go.

The teeth in this shark's mouth tell you how dangerous it can be.

As you study these animals, you will be amazed by how different they are from one another. Some, like the nudibranch (noo' dih bronk) pictured above, are amazingly colorful. Others, like the anglerfish lurking in the abyss (uh bis'), look a little frightening and even bizarre. Some, like the arrow squid, are absurd and brilliant. They hunt in packs and can leap into the air in an amazing display. Still others, like the dolphin and the manta ray, are majestic, intelligent creatures. Finally, creatures like the sharp-toothed sharks are menacing and frightful as they stealthily stalk the sea.

We will study many, many swimming creatures in this book, along with the fascinating things they do. Swimming animals are often called **aquatic animals**, because *aqua* is a Latin word that means, "water." Even though we know many facts about aquatic animals, there are still things about them that are mysterious. So much about them is still unknown to us because they are so hard to study, for their natural habitat spans the whole world! It's quite hard to follow, keep up with, stay near, film, photograph, and understand creatures that move about in such a vast environment. Over the years, however, scientists have been able to discover amazing things about the animals that live in the water. As you learn these facts, you will be filled with awe at the creativity of God and how many different kinds of marvelous creatures He made on the fifth day.

Have you ever wondered why God created so many different kinds of creatures that live in the water? Have you ever wondered why He created such wonderfully diverse creatures that do things we are just beginning to understand? I have, and I think I might understand the reason. I think God created all these glorious creatures because they delight Him. He enjoys and loves His creation. Even though the world is not the perfect, He still takes pleasure in the things He made. When people, who were created in the image of God, learn about these creatures, we can share in His joy and in the pleasure that God feels about the things He made.

Look at it this way: Have you ever done something that made you really proud? I have! And when I finished, I wanted to share it with the people I love the most — people who also love me. Did you feel that way when you accomplished something? You probably wanted others to share in the joy of your accomplishment. Well, that might be how God feels when we learn about His creation and all the wild and wonderful creatures He made. And, do you know what else? It brings glory to God when we study His creation and give Him credit for what He has done. It's not enough just to study science; we need to also acknowledge the Creator of it all. Let's glorify God this year by delighting in our studies of the creatures of the sea and giving glory to God, who made them all.

# Aqua Mobility

You might think that every animal that lives in the water can swim, but that is not the case! Some aquatic animals can only scoot or creep around, and many aquatic creatures can only float, moving wherever the water takes them. We call animals that can swim **nekton** (nek' tun), which comes from the Greek word that means "swimming." They get from one place to another by propelling, gliding, or paddling through the water. They usually have fins or flippers. Whales, seals, fish, sea snakes, turtles, octopuses, and squid are all **nektonic** (nek tahn' ik) animals.

This orange-lined triggerfish is a nektonic animal because it can use its fins, tail, and body to swim in the water.

This sea star (often called a starfish) is a benthic creature because it scoots across the ocean floor.

Animals that don't swim but scurry, crawl, hop, scoot, burrow, or slither across the bottom of a body of water are called **benthos** (ben' thahs), or **benthic** (ben' thik) **animals**. Even animals like sponges that attach themselves to the ocean bottom and don't move around are a part of the benthos. This word comes from a Greek word that means "depths of the sea." Can you think of an animal that might be benthic? Crabs, lobsters, sea snails, clams, and sea stars are examples of benthic creatures.

How can you tell if an animal is benthic? Is it always benthic if it sits on the bottom of a lake or ocean? No. Some fish, like flounder and stingrays, rest on the floor of the ocean for a long period of time, but they can also swim from one place to another. Because they can swim, they are nektonic animals. Benthic animals *cannot* swim from one place to another. Lobsters and crabs, for example, must walk across the bottom of the ocean. Because they are unable to freely swim about the ocean, they are benthic animals.

Some benthic animals, like sponges, are also **sessile** (ses' uhl). Corals are sessile, too. What do you think sessile means? It comes from the Latin word *sessilis*, which relates to sitting. Thus, sessile animals stick themselves to one place and just sit there. They don't move around. Sometimes, before they become sessile, these animals are **plankton** (plangk' tun).

You will hear a lot about plankton in this book, so let's talk about them. Their name comes from the Greek word *planktos*, which means to wander or drift, and that's exactly what plankton do — they drift. There are two kinds of plankton: **phytoplankton** (fye' toh plangk' tun) and **zooplankton** (zoh' uh plangk' tun). Phytoplankton are a lot like plants because they use the sun to make their own food. Zooplankton are more like animals. They need to eat to get food. In fact, zooplankton often eat phytoplankton! Although most zooplankton can swim a little, they are such weak swimmers that they cannot overcome the force of the currents. As a result, they drift to and fro at the whim of the waters.

While most plankton are very tiny, some are giant, like the lion's mane jellyfish, which rivals the blue whale as the longest creature on earth. The lion's mane jellyfish has tentacles that can grow to be over 100 feet long! Most plankton, however, are microscopic, which means they are so small you can't see them with your eyes. Instead, you need the help of a microscope to see them.

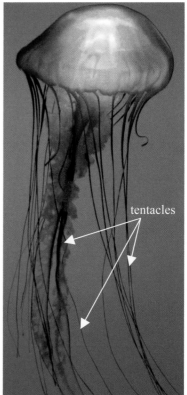

tentacles

Jellyfish like this one are zooplankton. They are such weak swimmers that they mostly drift with the currents.

You might be surprised to learn that most aquatic creatures are actually plankton when they hatch. For example, lobsters lay eggs that hatch little larvae that can't swim. Instead, they drift with the currents and are therefore plankton. Eventually, they grow into adult lobsters, at which point they are benthic animals. Most fish are also plankton when they are first hatched from their eggs. They cannot swim until they get older, so when they are young, they are at the mercy of ocean currents that go around and around the earth, and are carried this way and that way, wherever the currents may go.

It may seem like fun to ride the currents that circle the earth, but plankton have a truly difficult life. Without the ability to swim, they have very few ways to defend themselves from being eaten by other animals. And many animals, even giant ones like the 20-foot manta ray, the huge whale shark, and some great whales of the sea, eat plankton. They must eat tons of plankton each day just to survive. Where do you think these giant animals search to find plankton to eat? Yes, they search for currents. We'll learn about these currents later in this lesson.

Although it is huge (notice how big it is compared to the diver), this whale shark eats mostly plankton. Its mouth is designed to capture them as they drift with the currents.

Though ever so small, even microscopic, zooplankton that live in the oceans are truly phenomenal creatures. Can you imagine if every night you had to climb 25 miles up a mountain in order to get a bite to eat? Then, just when the sun began to creep up over the horizon, you had to climb back down so you would not become a meal for another animal. That would be a lot of work, wouldn't it? Yet that's exactly what life is like for many zooplankton. You see, most predators hunt during the day. So, the zooplankton sink deep into the ocean, hundreds of feet down, during the day to hide. Of course, all the phytoplankton that the zooplankton eat are near the surface of the ocean. Remember, phytoplankton must have sunlight so that they can make their own food, so they are found near the surface. In order to eat, then, the zooplankton have to climb back up to the ocean's surface under the cover of darkness to reach their food. Some are able to flap their tiny little fins or pump their little bodies enough to rise to the surface. This is an incredibly long climb for them, but they do it every single night in hopes of remaining alive long enough to grow into a larger creature. Now please understand that even though they can swim well enough to climb up through the water, they are not strong enough swimmers to overcome the currents, so they are still plankton.

As spring brings warmer temperatures and longer days, plankton can multiply so quickly that the water becomes cloudy with them — that's called a plankton bloom. Plankton blooms are most common in the arctic regions and happen each spring. Whales, dolphins, and hundreds of other animals paddle their way across thousands of miles to arrive for the plankton bloom. Many thousands of creatures feed upon plankton, depending on these small creatures for their very lives. In fact, if God did not create plankton, many of the animals I will discuss in this book would become extinct.

*You have learned about nekton, benthos, and plankton. Can you explain what you learned about them?*

# Filter Feeders

God created animals for many purposes and all magnificently display His glory. Yet some of these animals show us how practical and caring God is, such as the filter feeders He created. You will hear a lot about filter feeders in this book, because God made a lot of them. In every major animal group, there are usually one or two filter feeders. So what on earth are filter feeders? Filter feeders are animals that clean up the oceans and rivers of the world by eating the microscopic creatures and debris (duh bree') that float about in the water. Do you know what debris is? It is junk that has been discarded. Since they eat the debris, filter feeders are the "cleaning crew" of the water.

Water that goes into these filter feeders contains debris and microscopic creatures. This material is consumed and the "cleaned" water is expelled.

All the bodies of water on the earth — ponds, rivers, oceans, seas, and lakes — contain so many microscopic creatures and debris that without the filter feeders, the waters would become contaminated. Filter feeders can take in the contaminated water and spurt out clean water in its place. Some filter feeders are tiny, like small clams that live near the shore. Others are enormous, like the giant barrel sponge that filters many gallons of water each day. One thing is for sure: Without filter feeders, the bodies of water on earth would be full of junk and germs!

This strawberry sponge is an example of a barrel sponge. It filters thousands of gallons of ocean water every day.

# Animal Assortment

In this course, you will learn about a lot of animals, including **mammals**, **reptiles**, **amphibians**, **fishes**, and **invertebrates** (in vur' tuh brayts). Although you already learned some of these terms when you studied your first zoology course, I want to give you a brief overview right now to make sure you know what I am talking about.

Mammals are warm-blooded creatures that breathe air. Do you remember what "warm-blooded" means? It means their body temperature is always the same, no matter how cold it is outside. They also give birth to live young that drink milk from their mother's body. Mammals also have a backbone and hair. Some mammals, like whales, have hardly any hair, but they do have a few strands here and there.  You will begin your study of aquatic mammals in Lesson 2, when you learn about the largest of creatures, the whales.

Reptiles are cold-blooded creatures that have scales, breathe air, lay eggs, and have a backbone.  Do you remember what "cold-blooded" means?  It means that their body temperature changes with their surroundings.  Their bodies are warm when it is warm outside and cooler when it is cold outside.  Amphibians are like reptiles, but they don't have scales.  Fish are cold-blooded, have a backbone, and have scales like reptiles, but they don't breathe air.  They breathe under water using gills.

What do you think invertebrates are?  Well, a **vertebrate** (vur' tuh brayt) has a backbone.  When you add an "in" to the beginning, it means "without." So invertebrates are creatures that don't have a backbone.  We'll learn about invertebrates in the last part of this book.  You are likely to see many invertebrates when you visit the beach, including crabs, sand dollars, sea urchins, and sea snails.

# Current Events

Before we study the various creatures that live in water, let's learn a bit about the ocean, where many of these creatures live.  What helps whales migrate from the warm waters to the cold waters where the plankton blooms occur?  How can fish that feed on plankton find enough to eat?  The answers to these questions lie in a study of currents.  Currents are all about moving water from here to there.  The current in a river, for example, moves water from the beginning of the river (called the **head**) to its end (called the **mouth**).  Although it is easy to think of a current in a river, currents also exist in the ocean.  They carry cold water from the freezing areas near the polar regions (areas around the North and South Poles), along with millions of plankton, to warmer waters far away from the poles.  They also carry those warmer waters to the polar regions, which helps to even out the temperatures of the ocean.

Sea turtles often use ocean currents so that they can travel over great distances. If you saw the movie *Finding Nemo*, you already know about animals hitching a ride on currents.

Many sea creatures follow ocean currents along their winding paths up or down, across the ocean.  These creatures instinctively know where the currents are, hitching rides on them.  Other animals seek out the currents because they are places where food is found.  Huge amounts of plankton from the Arctic, for example, are caught up in currents and carried to other parts of the ocean.  Ocean currents, then, are like a giant food delivery system created by God to feed His animals!

# Surface Currents

Currents that form on the surface of the ocean are not surprisingly called **surface currents**. These currents are mainly formed by the winds. Interestingly enough, when you look at the surface currents in the world's oceans, you see that they form circular patterns called **gyres** (jires). To see what I mean, look at the drawing below:

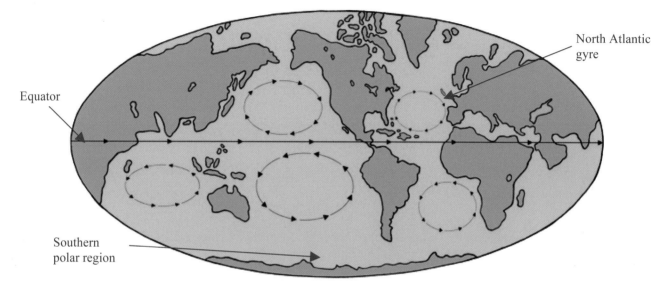

These are the major surface currents in the earth's oceans. The black arrows point the direction in which the currents flow.

Notice there is a current that pushes water straight along the equator, and another that pushes water straight along the southern polar region. The other currents, however, look like giant Ferris wheels that lie on their sides. These are the gyres. The gyres in the Northern Hemisphere run clockwise (the same direction a clock's hands turn), and the gyres in the Southern Hemisphere run counterclockwise (opposite of how a clock's hands turn). Notice the North Atlantic gyre pointed out in the drawing. It travels in a large circle from Florida up the east coast of the U.S., crosses over to northern Europe, travels down to Spain, and then goes across to Cuba and back up to Florida. To understand how winds are the major cause of these currents, do the following activity.

## Try This!

You will need another person, two hair dryers or personal-sized electric fans, some Cheerios or glitter, and a long casserole dish. Fill the casserole dish with water and sprinkle the Cheerios on the surface of the water. The other person needs to hold a hair dryer on one end of the dish, and you need to hold a hair dryer on the other end. Now turn both hair dryers on low and aim them just above the water. The other person's hair dryer should blow along one long side of the casserole dish, toward your end. Your hair dryer needs to blow along the other long side of the dish, toward the other person's end. Eventually, you should see the Cheerios start to flow in a circle. You have just created a miniature gyre! Just as your gyre was formed by winds from hair dryers blowing in opposite directions, ocean gyres are produced by winds on the earth that blow in opposite directions.

Here's an interesting story that illustrates how gyres work. One day, in the Pacific Ocean off the coast of California, a ship that was carrying Nike shoes sank. Thousands of Nike shoes were dumped into the ocean. Scientists predicted that because the shoes would be caught in the currents near where the ship sank, these shoes would wash up on beaches that lie along the clockwise gyre in the northern Pacific Ocean. Guess what? They did! They showed up on the beaches of California, then Hawaii, then the Philippines, and then Japan. Can you follow that pattern on the map on the previous page?

# Deep Ocean Currents

Some currents are not caused by the wind; they're caused by water temperature or the amount of salt in the water. Did you know that the deeper you go into the ocean, the colder the water is? You see, cooler water is heavier than warmer water, and so it usually sinks below the lighter, warmer water. Every summer, ice from the cold polar regions melts and cold water begins to flow out into the oceans. Because it is heavier than the warmer water at the surface, it sinks. It then moves slowly toward the equator, where it warms again and rises. This forms a large current flowing underneath the ocean.

A similar thing can happen when water evaporates from the surface of the ocean. When salt-water evaporates, it leaves the salt behind. The salt that's left behind makes the water on the surface of the ocean saltier, which also makes it heavier. This heavier water sinks to the bottom, forcing the lighter water to flow up to the top. Many times, it is both the temperature of the water *and* the amount of salt in it that causes the water to sink or rise, forming a deep ocean current. As a result, these currents are often called **thermohaline** (thur moh hay' line) **currents**, because thermo means, "heat" and haline refers to "salt."

*Tell someone in your own words what you have learned so far about filter feeders and currents.*

# Tides

If you have ever spent time at the beach, you may have noticed that the place where you set your stuff down when you arrived isn't on dry ground later on in the day. Every day, all day long, the water is either moving closer to the shore or farther away from the shore — back and forth it goes. These are the ocean's tides. When the water comes way up onto the shore, we call it **high tide**. When it pulls way back exposing a lot of the beach, we call it **low tide**. Many creatures are dependent upon the tides, especially animals that stay in tide pools.

These photos show the same beach at high tide (left) and low tide (right). Notice the tide pools that are formed at low tide.

Tide pools are created when the tide goes out, but crevices in rocks or the sand form pools of water. Some sea creatures get trapped in tide pools, while others make their permanent homes there. Those that get trapped wait for the tide to come in so they can slip out of the tide pool and return to where they normally live.

What causes these tides? Well, do you realize they are caused by an extraterrestrial (ek' struh tuh res' tree uhl) force — a force outside of this world? That's right! Tides are caused by the moon. It works like this: the moon pulls on the earth and its oceans with a force called gravity. As the moon pulls on the earth's oceans, the oceans bulge toward the moon, making an oval shape. At the same time, the earth is also pulled toward the moon, which makes the earth sit at the center of the oval.

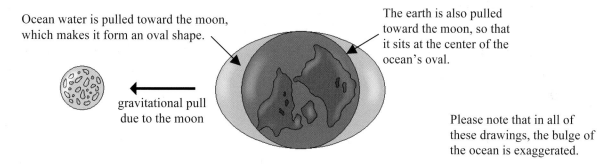

Ocean water is pulled toward the moon, which makes it form an oval shape.

The earth is also pulled toward the moon, so that it sits at the center of the ocean's oval.

gravitational pull due to the moon

Please note that in all of these drawings, the bulge of the ocean is exaggerated.

The moon takes more than 28 days to travel around the earth, so within the space of a day, it really doesn't move very much. Because of this, the oval formed by the ocean stays pretty much the same over the course of a day. The earth, on the other hand, spins completely around in one day. So think about what happens to a specific place (let's say a house) on the earth as the earth spins. In the diagram below, we are looking down on the North Pole of the earth:

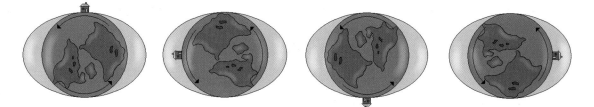

What happens to the house as the earth spins? At first, it is sitting on one of the flat sides of the ocean's oval, where there is not much water. As a result, it experiences low tide. As the earth spins, however, the house gets moved to the bulging side of the oval where there is a *lot* of water. At this time, then, the house experiences high tide. As the earth continues to spin, the house eventually gets to the other flat side of the oval, where it again experiences low tide. Eventually, however, the earth's spinning takes it to the other bulge on the oval where it once again experiences high tide. Over the course of the day, then, the house (really any place near the ocean) will experience a low tide, followed by a high tide, followed by another low tide, followed by another high tide.

Interestingly, the tides are not always the same. During a new moon or a full moon, the high tide is higher than usual and the low tide is lower than usual. We call these **spring tides**, even though they happen in all seasons, not just in the spring. Spring tides are caused by the sun working with the moon to pull on the oceans of the world. You see, the sun pulls on the earth and its oceans with its gravity as well. However, because the sun is so far from the earth, its effect on the oceans is small. Even so, when the sun's gravity adds to the moon's gravity, the high tides are noticeably higher and the low tides are noticeably lower. In the same way, depending on where the moon is compared to the sun, the sun's gravity can work *against* the moon's gravity, making the high tides lower and the low tides higher. When this happens, we say that the earth is experiencing **neap** (neep) **tides**.

*What can you tell me about currents and tides?*

# Planet Water

If you look down at the earth from space, you can see that the earth should have been named "ocean" instead of "earth," for the whole earth is mostly ocean! From up in space, it looks like a giant blue marble with white splotches. Why does it look blue? Because it has more water than land! Of all the animal habitats, the biggest and most abundant is the aquatic habitat.

Most of the waters of the world can be found in the oceans. Can you tell me how many oceans our world has? It has five: **Pacific**, **Atlantic**, **Indian**, **Southern**, and **Arctic**. Do you think you can find these oceans on a globe? Try that now. For another fun activity, look at how your globe is divided into a top half (Northern Hemisphere) and a bottom half (Southern Hemisphere). Which hemisphere is covered with the most ocean? Which is covered with the most land?

This is the earth as seen from space. Although you see some land (the brown areas), you see mostly ocean (the blue areas) and clouds (the white areas).

Connected to these oceans are several **seas**. Seas are smaller than oceans, but are made up of saltwater because they are connected to oceans. Although seas are mostly surrounded by land, they are usually joined to an ocean on at least one side. Can you find a sea on your globe? Look for a sea called the Mediterranean (med' ih tuh ray' nee un) Sea. It's just below Europe and above Africa. Can you find any others? The earth has lots of seas.

If you want to find an ocean, just follow a river in the direction of its flow, and you'll eventually get there, because the world's rivers and streams eventually flow into the seas or oceans. Where does the water that makes these rivers, lakes, and streams come from?  Well, rain is a big factor; however, ice melting on the tops of mountains, or underground springs that pour forth water each day also create rivers.  All these rivers are **freshwater** habitats, meaning they are not salty like the ocean.  When these waters reach the ocean, however, they become salty, or **brackish**.  The place where a river meets with an ocean or sea is called an **estuary** (es' choo air' ee).  Where the estuary is closer to the river, the water is less salty, and it becomes saltier the closer it is to the ocean.  Lots of creatures live in estuaries because food is plentiful there.  Although some animals can survive in both fresh and saltwater, most like a specific amount of salt in the water in which they live.  As a result, most creatures stay in a specific part of an estuary, where the amount of salt in the water is just right for them.

# Freshwater Facts

Have you ever wondered what the differences are among ponds, lakes, swamps, streams, and rivers?  Streams and rivers are made up of freshwater on the move — water continually flows through them.  Some rivers are wide, like the Colorado River, which has many narrow places, but can be as wide as a lake in other areas.  The beginning of a river is called its head, and the place where it empties into another body of water is called its mouth.  The water in a river is usually fairly pure at its head.  For example, many rivers begin in the mountains where melting ice is the river's source of water.  As the water flows down the river, however, it picks up all sorts of things like soil, bits of rock, and so forth.  If it is flowing by a polluted area, it could also pick up some pollution.  These things get carried by the river to its mouth, where they are dumped into another body of water.  Eventually, they will make it to a sea or one of the oceans.

Can you see which direction the water is flowing in this river?

Ponds and lakes are also freshwater habitats, but they don't have the rushing current that rivers have.  As a result, lakes are often dark and murky with the growth of plankton.  Many lakes are man-made, which means that people actually made them.  Usually, people make lakes by finding a river and plugging it up with a dam.  This causes the water to flood over a large amount of the land, making a lake.  The river still flows through holes in the dam, but only a little water is let out each day.  This way, the river still flows, but now there is a nice, big lake where there wasn't one before.  If the dam breaks, any town that popped up on the other side of the dam could be flooded and destroyed.

# Salt Solutions

You probably know that the water in the earth's oceans is salty, but did you know that most of the salt in ocean water is exactly the same stuff you sprinkle on your food? In fact, some brands of table salt come straight from the ocean. "But how does the salt get into the earth's oceans?" you might ask. One way salt gets into the oceans is by rivers that flow over rocks containing salt. As the river water flows over those rocks, it picks up some of the salt, and since the river water eventually ends up in one of the oceans, the salt ends up there as well. Another way salt gets into the earth's oceans is through volcanoes. Volcanoes that erupt under the oceans release salt into the water. Even salt that comes from volcanoes far away from an ocean can eventually make it into ocean water. In any case, the ocean is salty because salt is continually being added to it. Some parts of the earth's oceans are saltier than others based on how much salt is poured into that region.

*Can you explain what you have learned so far about the oceans and waterways of the world?*

# Creation Confirmation

Did you know that the oceans are getting saltier and saltier? This is because rivers and volcanoes dump salt into the ocean continually, but it is very hard for salt to leave the oceans. Because of this, the amount of salt in the oceans keeps rising. This actually tells us that the earth is not billions of years old, as some would have you believe. If the earth were really billions of years old, the amount of salt in the oceans would have been building up for billions of years, making them *much, much* saltier than they really are. In fact, the amount of salt in the oceans indicates that the earth's oceans (and the earth itself) are very young.

# Continental Shelf

Let's pretend it's possible for you to walk on the ocean floor without drowning. You start by walking out into the water from the shore. Obviously, you are not on dry ground, but it's still part of the continent (kon' tuh nent). It's called the **continental** (kon' tuh nent uhl) **shelf**. It is the part of the continent that is underwater. It slopes gradually downward, and the water

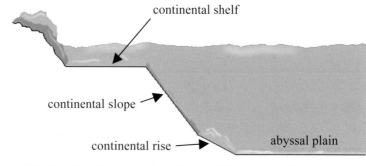

This drawing shows you how the land slopes away from a continent into an ocean. Please note that this drawing shows things much smoother than they really are.

gets deeper and deeper. Sometimes it's shallow for miles and miles, and sometimes it's only shallow for a few feet. When you get to the end of the continental shelf, there is a dropoff into the deep ocean below. This dropoff is called the **continental slope**. The continental slope is like a giant cliff, and like a cliff, there are places where it goes straight down and other places where it slopes down a bit more gently.  One part near the very end of the dropoff has a much more gentle slope.  This is called the **continental rise**.  The continental rise ends when it reaches the deep, dark ocean floor, which is called the **abyssal** (uh bis' uhl) **plain**.

As you walk farther and farther out on the continental shelf, you'll find the water getting deeper and deeper.  However, the continental shelf is usually not more than 600 feet deep.  Although 600-feet deep water is pretty deep, you are still on the continental shelf.  You may be out deep-sea fishing, but you're not fishing above the deep sea.  No, you're still fishing above the land that makes up the continent.  Sometimes the continental slope dropoff is not too far from the shore, such as in California; and sometimes the dropoff is far, far away from the shore, like in Siberia.

Along the continental shelf, you will find many different habitats such as mangrove forests, kelp forests, coral reefs, and sea grass meadows, and closer to the shore, rocky shores with animal-rich tide pools.  For further study, you can research each of these different habitats and learn where they occur and what kinds of animals you'll find in each habitat.

# The Abyss

If you were to leave the continental shelf and dive deeper and deeper down the continental slope, you would leave what is known as the **sunlit zone**, and enter the **twilight zone**.  While the sunlit zone is well lit by the sun, the twilight zone is fairly dark, with very little sunlight coming through.  A few animals can live in the twilight zone, but most prefer the sunlit waters above.  Long before you reach the bottom of the ocean, however, you will hit the **midnight zone**.  Why is it called the midnight zone?  Because it is as dark as — even darker than — midnight, for no light from the sun ever reaches down this deep into the ocean.

When you leave the continental slope, you reach the continental rise.  When that ends, you find yourself in the abyssal plain.  It is pitch black down here.  The sun's light is a distant memory.  Unlike the continental shelf, which may have hills and rocky areas with caves and caverns, the abyssal plain is pretty flat.  If you walked for thousands of miles, you would eventually run into huge mountains (the tallest mountains on the earth are at the bottom of the oceans), volcanoes, valleys, and deep trenches that span untold distances.  Most of the time, though, you would be traveling on flat land.

Most animals live on or in the waters above the continental shelf.  Very few live down in the deep ocean.  Can you guess why?  Think about what phytoplankton need to survive.  That's right!  They need the sun, because they make their own food using sunlight.  What feeds on phytoplankton?  Many creatures do, including zooplankton.  Also, other creatures feed on the creatures that feed on phytoplankton.  So, the animals tend to stay where the food is.  In the end, it turns out that most of the animals in the ocean live in the sunlit zone on the continental shelf and the waters above it.  Therefore, most of the animals in the ocean live right off the coast. Though the ocean is immeasurably enormous, covering most of the earth in water, most sea creatures live right next to us — near the land, near the shore, near the beach. God sure was nice to put most of the oceans' animals right near the shore so that we could discover and enjoy them!

# Abyssal Animals

The ocean floor in the deep, dark ocean is more immense than any land you have ever seen. It's bigger than any continent! It's so far down that no one has ever been to the bottom. Food is so hard to find there that most creatures live off very little, and few grow longer than several inches. While most animals live on the continental shelf or the waters above it, God did create special creatures to inhabit the vast abyssal habitat. Many of these animals have the ability to make their own light. This is called **bioluminescence** (by oh loo mih ness' ens). It is basically the same as the light that a firefly makes.

Since the deep, dark ocean is much like a huge, empty wasteland, what do the creatures that live there eat? Most eat dead animals that have fallen from the open ocean, animal feces that drop to the bottom, and bits of mucus (slimy waste from other creatures) they find floating. The animals that live here tend to have large mouths so that they can eat anything that happens to come their way.

Other animals that live here can attract things to eat. The deep-sea anglerfish, for example, has a built-in lure that it uses to attract other animals to it. The anglerfish lights its lure with bioluminescence, and then it wiggles the lure. When other creatures come to investigate this interesting source of light, the anglerfish eats them! It is not a very nice thing to do, but it allows the anglerfish to live in this deep, dark place. Do you know how this fish gets its name? Well, some people call fishermen "anglers." Since the deep-sea anglerfish catches fish with a lure just like fishermen do, it makes sense to use the word "angler" in its name!

bioluminescent lure

This deep sea anglerfish uses its bioluminescent "lure" to attract other animals so that it can eat them.

Besides deep-sea anglerfish, you can also find tiny white crabs, jellyfish, and gulper eels on the abyssal plain or in the deep water above it. Even though food is scarce for these animals, sometimes they happen on a real feast! For example, if a big whale dies and sinks to the ocean floor, animals on the abyssal plain and in the waters right above it will often find the whale. This "whale fall," as scientists call it, gives these animals enough food so that they feast for many weeks!

Now, we don't know a lot about the creatures that live on the abyssal plain and in the deep waters right above it. That's because we can't go down there to study them. If animals like deep-sea anglerfish can live down deep in the ocean, why can't we go down there to study them? We cannot go down there because of the **water pressure**. Do you know what water pressure is? Well, have you ever dived down to the bottom of a lake or pool? When you did that, did your ears pop? That was caused by the water pressing on you. Water is heavy, and when you pile a lot of water on top of your body, it begins to press on your body really hard. This is called water pressure. God designed the creatures that live in the deep ocean to handle this pressure. But if humans were to dive that deep, they would be crushed by all the water pressing down on them from above. Imagine lying on the ground and putting an enormous plastic bag on top of your body and filling it with water thousands of feet high. It would crush you just as if you dove down into the ocean and got underneath all that water.

Did you know that inside your body you have pockets of air? These pockets of air represent another problem when you try to dive into deep water. Think about your lungs, for example. Your lungs are like big air balloons inside your body. When you dive down, the water pressure is increased, and those "balloons" get scrunched smaller and smaller. When you come up closer to the surface, there is less water pressure, so your lungs grow larger and larger until they are back to the normal size they are meant to be when you reach the surface. Isn't that interesting? You can simulate this with a balloon.

## Try This!

For this activity, you will need a small balloon and a clear plastic 2-liter bottle (like a plastic soda pop bottle) with a lid. Hold the balloon in the bottle upside down so that the top of the balloon is in the bottle and the opening of the balloon is sticking out the opening of the bottle. Now blow up the balloon as much as you can. You won't be able to blow it up very big, because it will squeeze against the walls of the bottle. That's fine. Just do the best you can. Then, tie off the balloon to trap the air inside it. When the balloon is tied off, push it down into the bottle so that you now have a partially inflated balloon inside the bottle. Next, put the lid on the bottle tightly, so there is an airtight seal. Now lay the bottle on its side on the floor, keeping the balloon on one end of the bottle. While you are looking at the balloon, have a parent or older sibling step on the bottle (not where the balloon is) with all of his weight. The bottle should crumple where his foot is. What happens to the balloon? While you are still watching the balloon, have him lift his foot off the bottle. What happens?

When the person stepped on the bottle, he increased the air pressure in the bottle. What happened to the balloon as a result? It got smaller, didn't it? Stepping on the bottle simulates what happens to your lungs when you dive down into the depths of the ocean. As you dive deeper, the pressure on your body increases, and just like the balloon, your lungs get smaller. When he lifted his foot off the bottle, the pressure was relieved, and the balloon went back to its normal size. This is what happens to your lungs as you come back up to the surface. Now you can see one reason divers use oxygen tanks! Not only do those tanks provide the oxygen that the divers need to breathe, but they also increase the pressure in a diver's lungs so his lungs don't collapse when he goes deeper into the ocean.

# The Bottom Line

So, if scientists can't get down into the deep ocean, how do they know what is down there? Well, scientists have designed scuba diving suits that are pressurized, allowing scuba divers to get over

1,200 feet deep, but that's not anywhere near the abyssal plain. To see down that far, God has enabled man to design large machines, sort of like spaceships that go under water instead of into space. These "underwater spaceships" are called **submersibles** (sub mur' suh bulz). Some submersibles carry people, and some are unmanned, meaning that they go down without a person inside. Unmanned submersibles are less expensive to build and are able to take pictures of what is down under the ocean and bring back samples of what is found down there. However, scientists like the manned submersibles better, because they want to actually go and see for themselves!

This submersible, called "Alvin," carries three scientists and has been to depths of more than 6,700 feet.

Now that you have learned a lot about the places where the aquatic animals live, the rest of this book will focus on the animals that dwell in these places. Before you go on, however, it's time to spend some time reviewing what you have learned and experimenting so that you can learn a bit more.

# What Do You Remember?

What are nektonic creatures? What are benthic creatures? What are plankton? Where can zooplankton be found at night? Why are plankton important to all sea life? What are filter feeders? Can you name the five oceans in the world? What are seas? What are estuaries? Beginning from the shore out to the deep, what are the four zones of the ocean floor? From the surface of the ocean to the deep, what are the three zones in which aquatic creatures live? What are the circular currents called? What are the currents caused by temperature and salt levels called? What causes the tides?

## By the Beach

If you can go to a beach, the best place to look for sea life is in tide pools. Tide pools are filled with all manner of sea life. They are best found on rocky shores, but even a sandy shore can have tide pools. You can often find sea life hidden under rocks and in crevices in the tide pools. Look carefully, and you might discover a wonderful world of aquatic animals!

This tide pool was formed when the tide went out. It is a great place to look for aquatic life.

## Freshwater Finds

If you can go to a lake, look for signs of animal life in the water.  Do you see unusual clumps of mud possibly made by an aquatic creature?  Do you see the telltale marks of a slithery water snail?  If you look carefully, you are sure to see signs of life.  You might even see aquatic creatures for yourself.

# Your Notebook

It is important that you review this material before you move on to the next lesson.  You see, as a student, it isn't enough to just read and learn.  You need to put information on paper by drawing pictures (illustrations) and writing (or dictating) what you have learned.  This will help you to remember it longer, and it will provide evidence of what you learned.  The main way you will review material in this course is to make a notebook of your zoology studies.  You will make illustrations, do fun assignments, record all that you learn, keep "Scientific Speculation Sheets" from experiments, and even add pictures of other things you see and do.  Your notebook will be a collection of your zoology studies.  When you look back over it in the years to come, you will be reminded of the many sea creatures and fascinating facts that you learned in this study of aquatic creatures.  You can create your own notebook or use the *Zoology 2 Notebooking Journal* to do all your assignments.

If you haven't already done so, record in your notebook some of the fascinating facts from this lesson. Also, make drawings that explain what causes the tides and what the regions of the ocean floor are called. If you have the *Zoology 2 Notebooking Journal*, pages are provided for you.

# Ocean Box

You are going to create a box to display your own models of the animals you learn about in this book.  Today you will build this box called — your **ocean box**.  You can choose any size box — a small shoebox or a large shipping box.  You need to line it with blue paper, just like the boy is doing in the picture on the right.  That way, it looks like the ocean.

In each lesson you will learn about some aquatic animals, and you will then make a model of those animals and add them to your box.  You can either use clay or you can use pictures from magazines or printed from the Internet.  At the end of the course, you will have a box filled with sea life like the one shown on the left.  Most of the animals in this ocean box were created from clay and either glued to the box, stuck to the box with tape, or hung from the top.  As you put your ocean box together, do a good job so that you can be proud of your accomplishment when you are done with this course!

# Experiment

We discussed currents caused by heavier water sinking below lighter water. The question I would like to ask is: Do you remember which is heavier, cold water or hot water? Let's do an experiment to find out.

You will need:

- A "Scientific Speculation Sheet" (found on page iv)
- A large clear bowl or container (It could even be the bottom half of a plastic soda bottle.)
- A paper or Styrofoam cup
- A nail or pen to puncture a hole in the cup
- Blue and yellow food coloring
- A spoon with which to stir
- Hot water and ice-cold water

1. On your "Scientific Speculation Sheet," record your hypothesis about which will be heavier: hot water or cold water.
2. Fill a large glass bowl with hot water. You can warm the water in the microwave if you wish. **Make sure it is not too hot!**
3. Place a drop or two of blue food coloring in the water and stir.
4. Pour ice-cold water into the cup and add several drops of yellow food coloring. Stir so that the ice water is yellow.
5. Holding the cup over the sink, puncture a hole in the bottom of the cup with a nail or pen.
6. Hold your finger over the hole and slowly place the cup in the hot, blue water.
7. Pull your finger off the hole in the bottom. What happens? Why do you suppose the cold water pours into the bowl? Does it appear to be mixing well? Which way is it moving as it pours into the bowl?
8. Now let's change the experiment. Empty the bowl and this time fill it with ice-cold water, adding a few ice cubes to keep it cold.
9. Add a drop or two of blue food coloring and stir.
10. Place your finger over the hole and fill the cup with hot (**not** *too* **hot!**) water, and add yellow food coloring to the hot water. Stir to make sure the food coloring mixes with the water.
11. Place the cup in the bowl of blue ice water and take your finger off the hole. What happens? How is this different from what you saw in step 7? Why is this happening?
12. Based on this experiment, which is heavier: hot water or cold water?
13. Fill out the rest of the "Scientific Speculation Sheet" and place it in your notebook.

# Lesson 2
# Whales

What animal has a tongue that weighs more than most cars, a heart as big as a Volkswagen Beetle, blood vessels so large that a baby could crawl through them, and spends most of its life in parts of the ocean that no man has ever seen? Why, it's the largest of God's creatures – the **blue whale**. It may surprise you to know that the blue whale, though it's the biggest creature in the entire world, is so hard to find that we know very, very little about it. In fact blue whales are among the few sea

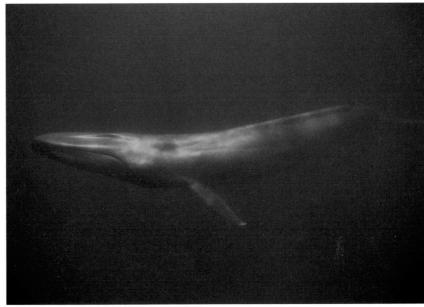

Blue whales like this one are very hard to find, even though they are the biggest animals in creation.

creatures that don't stay near the shore – they spend most of their lives out at sea – traveling thousands and thousands of miles between the waters in which they eat (their feeding waters) and the waters in which they give birth to their young (their breeding waters).

Even though a whale can dive thousands of feet below the surface of the ocean, it must come up to breathe every now and again, or it will drown. Why? Because whales are mammals, and like you and me, they breathe oxygen from the air. Unlike many sea creatures, whales cannot breathe under water. As a result, they must come to the surface regularly so they can breathe the air. Mammals that live in the sea are often called **marine mammals**, because the word "marine" is used to refer to the sea.

## Two Kinds of Whales

Whales are called **cetaceans** (see tay' shuhns), because they belong in Kingdom **Animalia**, phylum **Chordata**, class **Mammalia**, and order **Cetacea** (see tay' shah). Do you remember the taxonomy you learned about in your first zoology course? If not, you might want to review the first lesson in that book, because I will talk about it from time to time in this book.

To help you remember the word "cetacean," I'll use it a lot in this lesson. Cetaceans are divided into two different kinds: **baleen** (buh leen') **whales** and **toothed whales**. A baleen whale has no teeth. Instead, it has long strips of bristled plates that hang from its upper mouth, like an enormous toothbrush permanently attached to its upper lip. This structure is called baleen. Baleen whales are also called great whales – not because they are any greater than other whales, but because they are usually much, much bigger than toothed whales.  In fact, when you think of the word whale, you probably imagine a huge creature.  But did you know that a dolphin is a kind of whale too?  Dolphins are toothed whales.  And guess what toothed whales have instead of baleen?  Of course!  They have teeth!

All whales, both baleen whales and toothed whales, have big brains.  It's no wonder they are considered the most intelligent of all marine mammals.  The dolphin is so intelligent, for example, it actually plans ways to catch food that only a very skilled hunter could invent.

# A Whale of a Tail

The end of a whale's tail is called a **fluke**.  The whale uses its fluke to steer when it is moving through the water.  It also uses its tail for power when it swims.  Interestingly enough, whales move their tails up and down in order to power their swimming.  What's so interesting about that?  Think of how a fish swims.  It swims by moving its tail from side to side.  Whales, on the other hand, move their tails *up and down* in order to swim.  So the way a whale swims is quite different from the way a fish swims.  Of course, you already know

This is the fluke of a humpback whale.  The whale moves its tail up and down in order to swim.

another big difference between fishes and whales, don't you?  A whale cannot breathe under water, but a fish can.

# Do You Hear What I Hear?

Cetaceans can't smell very well, if at all, so they really depend on their others senses, like seeing and hearing, for finding each other and food.  Interestingly enough, seeing is not the most important sense for cetaceans.  Hearing is.

If you were on the other side of your neighborhood and your mom walked outside of your house to call your name, would you be able to hear her?  What if she was calling your name from the parking lot of the grocery store down the street?  Could you hear her?  Probably not.  Well, sound travels better under water than it does in air, so cetaceans can hear sounds hundreds of miles away.  In fact, some cetaceans seem to be able to communicate with other cetaceans that are thousands of miles away!  Isn't that amazing?

Every cetacean has its own kind of sounds. To us, they might sound like clicks, whistles, moans, or rumbles. They talk to one another a lot, especially when they are in a group swimming under the water. The most famous is the male humpback whale that sings a low, moaning song which usually lasts for 10 to 20 minutes. The male repeats this song many times each day. There are parts of the song that we cannot hear with our ears, because our ears can't hear the same range of sounds that a whale's ears can. Nevertheless, scientific instruments can be used to collect and analyze those parts of the song. Humpback whale songs can travel hundreds of miles, allowing them to communicate with groups of whales that are very far away.

All of the male humpback whales in a given group sing the same song, but that song does change over the course of a season. Because biologists don't fully understand whales, they are not certain why humpbacks sing so much, but many think that the males sing in order to find a mate. If you visit the course website I told you about in the introduction to this book, you can listen to some whale songs that scientists have recorded.

*Explain all that you have learned so far about whales. Be sure to include information about their tails and their senses.*

## Thar She Blows!

spout →

whale's back

The spout of water you see in this picture is coming from the whale's blowhole. As the whale surfaces, it exhales through its blowhole, and the water in the exhaled air turns to steam.

As I mentioned before, cetaceans must breathe air. Typically, a cetacean will breathe through its "nose." You and I can also breathe through our noses, can't we? But where is a cetacean's nose? If you search and search the front of the cetacean's face, you won't find a nose anywhere. So how does this creature breathe? Well, a whale has a "nose" on top of its head! It is called a **blowhole**. When the whale goes under the water, it can close its blowhole so water can't get in. When the whale surfaces, it opens its blowhole and exhales. As this happens, a spout of water vapor rises into the air. This gives people a way to look for whales when they surface to breathe.

Typically, a whale will take in one breath after it reaches the surface and exhales, then it will go back down under water until it is time to breathe again. Remember, once the whale goes under water, it closes its blowhole so that no water can get in. This breathing system is perfectly designed by God for the whale!

People who hunted whales in the early days used to say, "Thar she blows!" when a whale surfaced and blew a spout into the air. They could even tell what kind of whale it was by the size, shape, and height of the spout. Some whales, like the blue whale, have a spout that goes up more than 30 feet. It would be as high as five or more men standing on top of one another! A sperm whale's spout could reach 15 feet. That would be like piling almost three men on top of each other. Interestingly enough, although toothed whales have only one blowhole, baleen whales usually have two.

The spouts that a whale makes when it exhales make it look like the whale is blowing out water, but it's not. It's blowing out air! However, because the air was warmed up so much inside the whale's body, and because the air above the ocean is cooler than the whale's body, the warm air turns to steam when it hits the cooler air. So, it looks like the whale is blowing water out of its blowhole, but it is really just warm air that turns to steam. You've probably done this yourself on a cold day, haven't you? When it is really cold outside, you can "see your breath," because the air you exhale is so warm compared to the outside temperature, it turns to steam as it leaves your body.

The whale's breathing system is amazing, and it is even more amazing to realize that it is completely separate from the whale's mouth. Why? Well, have you ever taken a drink and ended up coughing uncontrollably? If you have, it was because when you swallowed, some of what you were drinking got into the tube that leads to your lungs. God designed you so that your throat leads to *both* your stomach *and* your lungs. At the point where your throat ends, it splits into two tubes: your **esophagus** (ih sof' uh gus), which leads to your stomach, and your **trachea** (tray' kee uh), which leads to your lungs. When you eat or drink, a small flap, called your **epiglottis** (ep' uh glot' is) covers your trachea, keeping food and liquid from entering your lungs. However, every once in a while the epiglottis doesn't close fully, and liquid or food gets into your trachea, and you automatically cough to try to get it out. That's one reason you can't really eat under water. If you took in water with every bite, there would be more chances for it to get past your epiglottis and into your trachea.

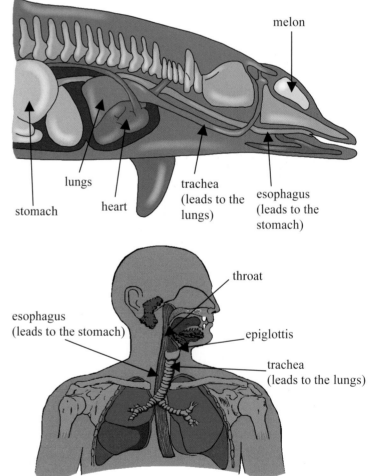

In the whale (top), the lungs are connected to the blowhole by a different tube (the trachea) from the one that connects the mouth to the stomach (the esophagus). In people, the trachea and esophagus are both connected to the mouth.

A whale, with its blowhole on top of its head, has a separate tube leading from its mouth to its stomach. This tube is not connected to its lungs at all. A whale's lungs and mouth have a totally different design from any other mammal. So, there is no danger of choking and drowning when it opens its mouth under the water to eat and swallow food. The water can only get into its lungs through the blowhole, which it closes tight when it is under the water. The only way a whale can drown is if it is prevented from reaching the surface of the water for a breath of air. This can happen if a whale is attacked by killer whales, which hold young whales down under the water to drown them.

Since a whale must come to the surface to breathe, you might wonder how it sleeps. Scientists aren't 100% sure. However, they have noticed that whales do take short naps (usually for less than an hour) at the surface of the water where they can breathe. In addition, some whales (like dolphins) can actually put some parts of the body to sleep, while the rest of the body stays awake so that the dolphin can swim to the surface to breathe.

## Beach Bum

Sometimes, whales get stranded on the beach. When this happens, we say the whale is **beached**. Beached whales are helpless because they are not designed to come out of the water. Often, if rescuers don't find them and help them back into the water in time, they will die. Strangely enough, when a beached whale is found in time, and rescuers have spent hours carefully moving the creature back into safe waters, it will sometimes just beach itself again. Why do whales do this? We really don't know. Some scientists think whales beach themselves because they are disoriented or sick. Even more strange, sometimes entire *groups* of cetaceans beach themselves!

This blue whale died because it was beached.

Once, when I was a child, my parents took me to the beach in Corpus Christi, Texas. When we arrived, we saw a dozen huge whales washed up on the beach. There they were, all these whales, dying on the beach. It was a very sad thing to see. How could an entire group of whales get beached? Once again, scientists don't really know. Some suggest that the whales in a group are so loyal to each other that when one becomes disoriented due to illness or injury and swims up on the beach, the others just follow.

## Whale Moves

Whale watching is an activity that many people enjoy. Whales almost seem to perform when people come to watch them. They flip and flop and smash their tails on the surface of the water. Of course, they aren't really performing. They are just doing what they do naturally.

One whale move that delights whale watchers is called **breaching**. This is when a whale leaps into the air and then purposefully flops down on the water with an incredible splash. Sometimes it twirls in the air when it does this. Scientists aren't sure whether breaching is done for play, to clean the whale's skin of things that are stuck to it, or to communicate something.

Another interesting whale move is called **spyhopping**. This is when a whale sits straight up in the water with its head positioned straight up and out of the water. It will sometimes turn around in circles as it spyhops. Scientists think this might be an effort to check out the surroundings above water.

Although you can often see whales breaching like this in the ocean, scientists aren't sure why they do it.

**Lobtailing** is done when a whale faces downward in the water with only its fluke sticking out. It then slaps the water with a thunderous sound. Scientists think this could be done to warn other whales of danger, but they aren't sure, because it isn't consistent. **Logging** is yet another whale move. This is when a whale swims slowly at the surface of the ocean with very little movement. When a whale does this, it looks like a log in the water. Some scientists think that this is a form of rest or sleep for whales.

Though many thousands of people enjoy whale watching and find joy in the awe-inspiring antics of these great giants, it hasn't always been this way. You see, many people used to hunt whales, not to observe them, but to harm them. Let's explore the reason for this.

# Whalers

Cetaceans have a thick layer of fat under their skin called **blubber** (blub' ur). In some species, the blubber can be two feet thick. That's a lot of blubber! For many years, the Eskimos of Alaska depended on the whales for their very lives. Every part of a whale was used - the bones, the blubber, and the skin – every bit. In this way, one whale could supply the Eskimo community for a long time. However, sometimes a good thing can get out of control and become a bad thing. This happened with whale hunting.

You see, whale blubber was not only used by the Eskimos. It was once an important product that every person in the civilized world wanted. This is because, way back before electricity was discovered, blubber was used as oil for lamps. If you wanted to see what you were doing after the sun went down, you needed to light lamps, and you needed oil to do this. The oil from whale blubber had other uses as well.  It was used to oil machines, for example.  In addition, a smelly, gooey substance found in the intestines of a particular whale (the sperm whale) was used to make extremely expensive perfume.  In the early 1900s, one sperm whale's intestines might contain several thousand dollars worth of this goop.  Therefore, whales were hunted and killed so that people could make money.

The fishermen that hunted whales were called **whalers**.  As time went on, whalers figured out that whales tend to travel along the same paths in the oceans during certain times of the year.  You'll learn about that in a little while.  Once whalers figured this out, they followed the whales as they traveled and hunted them mercilessly.

Though it brought in a lot of money, whaling was an incredibly dangerous and frightening job.  In the early days of whaling, people sat in watchtowers and alerted whalers when a whale was spotted close to shore.  Then the whalers set out in small boats to chase it down.  When they got close enough, they threw heavy, sharp harpoons into the body of the whale.  The harpoons were attached to ropes that were attached to the boat.  Harpoon after harpoon was hurled into the poor whale.

Whales that were harpooned would give off warning cries that could be heard by other whales hundreds of miles away.  This cautioned other whales to stay far away from that area.  Some whales were able to fight back.  They might pull a boat furiously through the water at breakneck speed.  Other times, a hurt whale could overturn a boat by bumping it hard enough.  Usually, though, the whalers would wear down the whale.  Soon, hurt and exhausted, the whale did not have the strength to resist.  It was then towed back to shore, where it was cut open and the blubber, intestinal substances, and other resources harvested and sold.

This drawing depicts whalers from the past doing their job.

Larger and larger boats and better equipment for hunting whales were eventually built.  Soon whales had no chance of escape.  Huge whaling ships could harvest many whales on a single trip to sea.  In the later days of whaling, whales were hunted way out at sea.  Sometimes entire whale families were found, and as many were killed as possible.  Sadly, this process caused the near extinction of many whales.  So now whales are protected by **conservation** (kon' sur vay' shun) **laws**.

Conservation simply means saving something.  If you conserve money, you don't spend it; you save it.  If you conserve whales, you save whales by keeping people from killing too many of them.  Conservation laws, then, help save whales.  Also, since chemists have developed many products that take the place of whale blubber, we don't need to kill whales anymore.  So, the scientists who made these products have saved cetaceans from extinction!  Maybe one day you will be a scientist who discovers a substance that will save an animal from extinction.

*Can you name the moves that a whale makes?  What else have you learned so far?*

# Migration

Many cetaceans migrate incredible distances every year.  They often summer in cool waters like those in the polar regions (areas near the North or South Pole), where there's plenty of food.  So every summer, many thousands of cetaceans head to the cool waters, eating tons of food each day, which increases their stores of blubber.  They'll need this extra blubber for the winter, when they head to the warmer waters near the equator to have their young.  You see, in these tropical waters, there isn't a lot of food that whales like, so they generally don't eat the entire time they are there, which can be months.  After they have their young and the weather begins to show signs of spring, they head back to the polar regions to eat again.

Every year, it's the same routine – head to the polar regions for the summer and towards the equator for the winter.  Most whales make their trips in groups, called **pods** or **herds**.  Some mother whales that are bringing their newborns to their feeding grounds must travel alone because the calf is too slow to keep up with the pod.

*Explain all that you have learned about how a whale breathes, whale beachings, and migration.*

# Don't Have a Calf

Like all mammals, whales give birth to live young, called **calves**.  They don't lay eggs.  Animals that give birth to live young are called **viviparous** (vye vip' ur us) animals.  Usually the calf is born tail first.  Once it is born, the mother (or another adult) immediately guides it to the surface for a breath of air.  Soon, the calf is adjusted and ready to drink its mother's milk.  The mother's milk glands are hidden in a little pocket of skin near the mother's tail.

This sperm whale calf is swimming alongside its mother.

The calf latches onto the mother and drinks as she swims about the ocean.  Whales are great mothers, caring for their young every waking moment.  They become very attached to their calves, helping them swim, breathe, and grow up to be a big whale.

Soon after the calf is born, the mother and her young join other whales on their journey back to their feeding grounds. It's a long way. When you were small and went places where there was a lot of walking, did you ever ride "piggy back" on your mom or dad so you could rest? Often the calf tires and must be held up above the water on the mother's back as she continues the journey. It's like a piggy back ride for the calf.

On the migration back up to the feeding grounds, the mother doesn't eat on the entire journey, but the calf will drink milk from its mother the entire time. This weakens the mother, who is living off the food she ate the summer before. This makes the journey to colder water a slow trip for the mother and her calf. Sometimes the mother and her calf can't keep up with the pod and are left to make the journey alone. This is a dangerous trip for them, for killer whales are looking for just such an opportunity. When killer whales see a mother and its calf, they are sure the calf will be easy prey, for a mother can only defend its calf for so long before it tires of the fight. If the mother and calf make it back to the feeding grounds, they'll be safe, because the food will restore their energy and they'll have the company and protection of the pod.

The rest of this lesson will focus on each of the different kinds of whales. Keep reading if you are curious about creatures like porpoises, dolphins, killer whales, humpback whales, and unicorns. Wait a minute…did I say unicorns? Yes, I did. To find out what I mean, you'll have to read on.

# Toothed Whales

Do you remember that there are two kinds of whales: toothed whales and baleen whales? Well, we are going to take a moment to learn about a few of God's beautiful and delightful toothed whales, and then we'll discuss baleen whales.

Do you remember what makes a toothed whale a toothed whale? Did you say teeth? You're right of course! As I mentioned, some toothed whales have a few teeth, some have several hundred. These teeth look like upside down ice cream cones – we call them cone shaped.

Notice how this whale's teeth look like upside down ice cream cones.

But whales don't use these teeth to chew, because they don't chew. They just swallow. A whale uses its teeth to hold its prey while preparing to swallow it. Smaller prey are tossed into the air and swallowed whole. You can watch this happen at large aquarium shows when animal trainers toss fish to dolphins or killer whales.

# Echoes to Locate

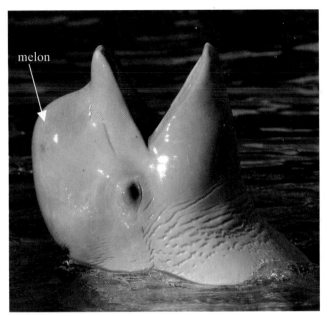

melon

This beluga whale has a very pronounced melon. To see where the melon is on the inside of the whale, look at the drawing on page 22.

Most toothed whales have a **melon**. No, they don't have a piece of fruit. What they have is a protruding, rounded forehead that contains a structure called a melon. It is used for **echolocation** (ek' oh loh kay' shun). Do you remember learning about echolocation when you studied bats? Well, just as bats use echolocation to "see" in the dark, toothed whales use echolocation to "see" under water. Scientists believe toothed whales are the only whales that have this ability. We sometimes call echolocation **sonar**, because scientists have made an echolocation system for ships and submarines that is called sonar. Even though ships and submarines use manmade sonar, whales use God-made sonar – which is far better!

How does echolocation work? Well, toothed whales make special sounds that are sent out through the melon. The sounds bounce off things in the sea and come back to the whale. God designed parts of the toothed whale's body to receive the returning sound, which is called an **echo**. God also gave the whale's brain the ability to use the echoes to determine the size, shape, location, and composition of the object very precisely. It's an amazing ability! The toothed whale can tell if there is a school of fish up ahead, or a wall, or a boat. It knows how far down the ocean floor is, how far away the shore is, and even if sharks are nearby. This special gift of echolocation enables dolphins to know where they are going and to "see" what is all around them.

## Try This!

Roll two pieces of cardstock or poster paper (8½ x 11 works well) into two megaphones. Use tape to keep each of the megaphones from unraveling. Have one person talk through the megaphone towards a wall in your house. At the same time, have another person place the small end of the other megaphone next to his ear, standing at a slight angle to the wall, listening. Try listening both with and without the megaphone. What happens? Some objects reflect sounds well, while others don't. Try this on different objects, like a metal refrigerator, a wooden cabinet, and a brick wall. Do you notice a difference? Can you see how a toothed whale uses sound to determine what kind of object is near?

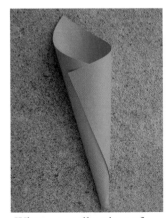

When you roll a piece of paper into a cone, you can use it as a megaphone.

# Dolphins

There are a lot of different kinds of toothed whales, and I want to talk about several of them. I'll start with what I think is the most delightful animal in the ocean – the dolphin. Dolphins have a long history of interacting with humans. People love them because they are playful, intelligent, and always seem to be smiling. Many stories told in both ancient and recent history talk about dolphins saving people from drowning at sea and bringing them safely to shore. Scientists aren't sure whether dolphins do this because they actually want to save a person or because of instinct.

The bottlenose dolphins in this pod look after one another.

Why do scientists think dolphins save drowning people by instinct? Well, if a dolphin behaves like it might be drowning, another dolphin will push it up to the surface of the water so it can breathe. An injured dolphin is hoisted on the back of a fellow dolphin and carried to shallow waters where it is able to breathe and recover. Since dolphins do this to their drowning dolphin friends, it would seem natural that they would know how to save a drowning person as well. So, do you think a dolphin saves a man from drowning because it is an instinct, or because it purposefully wants to save the drowning man?

Dolphins swim in pods; an extremely large pod is called a **herd**. They are very social and help one another fight off predators, like sharks. They can kill large sharks by ramming them over and over again with their pointed beaks and their melons. They even babysit for the moms that need to leave their calves to hunt for food.

Dolphins are extremely intelligent, perhaps some of the most intelligent animals that God created. They prepare and carry out complex plans for trapping food. For example, they often hunt for halibut, a type of fish that lies on the sand at the bottom of the ocean. To hunt these fish, dolphins will line up in shallow water, and one member of the pod will disturb the halibut resting on the sand. The startled fish will jump out of the sand – right into the line of waiting dolphins' mouths! A pod of dolphins will sometimes circle a school of fish, continuously swimming around them to form a barrier that keeps the fish from escaping. Then one by one, the dolphins pick off each fish in the circle. Dolphins may also send out high-pitched sonar sounds that stun prey so it can then be gobbled up.

There are more than thirty different species of dolphins, and they live in a lot of different places: in the deep sea, near the coasts, and even in some rivers. They have bodies that are shaped like bullets, which allow them to swim very quickly. Even though they may weigh more than 1,000 pounds, they can leap right out of the water and into the air. When they race along near the surface and make low leaps out of the water, it is called **porpoising** (por' pus ing).

This dolphin is porpoising.

The bottlenose dolphin is the most commonly seen dolphin. It tends to stay near the coast in the warmer Atlantic waters. There can be a few hundred of them in one pod. It is also the most easily tamed dolphin and is often found in marine exhibits, like Sea World. It is gray on top and white on the bottom.

# Porpoises

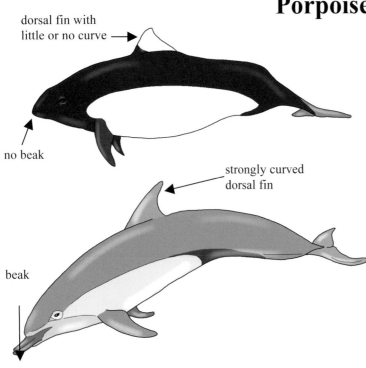

dorsal fin with little or no curve →

no beak

strongly curved dorsal fin

beak

These two drawings illustrate the differences between a porpoise (top) and a dolphin (bottom).

Even though dolphins can engage in an activity called porpoising, there are differences between dolphins and **porpoises** (por' pus ez). In general, porpoises are smaller than dolphins. Porpoises, for example, rarely grow longer than 7 feet, while dolphins can grow longer than 10 feet. In addition, dolphins usually have a lean, sleek body, while porpoises tend to look "chubby" by comparison.

There are other ways you can tell porpoises and dolphins apart. Dolphins have beaks, while porpoises do not. A dolphin's beak is not like a bird's beak, but it is something you can identify, as shown in the drawing on the left. Many pictures not drawn by scientists will inaccurately show a porpoise with a beak. If it has a beak, it's probably not a porpoise. Another difference between dolphins and porpoises is the shape of the dorsal fin. A dolphin's dorsal fin curves, while a porpoise's is triangular with little or no curve in it.

# Killer Whales

Have you ever seen an **orca** (or' kah)?  If you have seen a **killer whale**, then you've seen an orca, because they are the same animal.  Orca sounds nicer than killer whale, though, doesn't it?  If you have been to a large aquarium like Sea World, you have probably seen these friendly whales carrying passengers, doing tricks, and seeming unlike killers.  Why, then, are they called killer whales?  Well, these animals, which can reach lengths of over 30 feet, are considered the most ferocious of all

the animals in the sea.  Packs of fifty or more, all working together, surround and chase sharks, seals, dolphins, penguins, other whales – especially weak nursing calves – and any other hapless creatures they find innocently swimming along.  First, they all chase the creature, making it tired and weak.  Then one of them will move in to take a bite, with all the others quickly following.  They are like the "sharks" of the whale world, but most animals fear killer whales even more than sharks!  These whales have been known to kill prey even though they weren't hungry enough to eat it.  Sometimes, they hunt an animal, ferociously kill it, and then just let it drop to the bottom of the ocean.

This picture shows you the colors and markings of killer whales.

The dorsal fin of the orca can extend up to six feet above its body.  That's taller than most grown men!  And because a killer whale swims close to the surface, the dorsal fin can often be seen gliding through the surface of the water.  This causes some people to mistake killer whales for sharks.  You can usually tell killer whales by their markings.  They are mostly black, with white markings on different parts of the body.  It is possible, however, to mistake certain porpoises, called Dall's porpoises, for killer whales, because they have similar markings.

Killer whales, though ferocious to sharks and other creatures, are generally docile (dah' sil) and friendly towards people when kept in captivity.  They are often tamed and can be trained to do tricks for audiences in marine shows.  Extremely intelligent animals, killer whales have even starred in movies, such as the film titled *Free Willy*.

# Beluga Whales

In the cold, icy waters off the coast of Alaska lives a whale known as the **sea canary**.  Like the canaries that fly in the sky, these whales make all sorts of tweeting, squeaking, and chirping noises which can be heard above and below the water.  Although we think they do this for communication, we aren't really sure. Regardless of why they make these noises, they are a delight to individuals who witness the spectacle. These beautiful, white whales are simply called **beluga** (buh loo' guh) **whales**. The word beluga means "white one" in Russian.

Beluga whales are playful and generally friendly towards people.

Beluga whales are beautiful to see, and at first glance, they look like fat, white dolphins, except they don't have a beak. Also, they grow to weigh more than 3,000 pounds. They act a lot like dolphins as well, being very friendly towards people.

A beluga whale has a big melon on the top of its head, which is used for echolocation (see the picture at the top of page 28). It also has a smooth, finless back. A pod of these whales might have a hundred in all, ranging in size from 15 to 18 feet long. The young pups are born brown or black, usually turn blue by the time they are one year old, eventually turn light yellow, and then become snow white as they mature.

Every year, the belugas travel in large pods up through Alaskan estuaries to extremely shallow waters in order to **molt**. When they molt, the outer layer of the skin peels off, revealing new white skin. In the shallow waters, they rub their bodies against the gravel floor to help get rid of the old skin.

Unfortunately, beluga whales have been known to stay too long in the north, getting trapped under the moving ice sheets that begin to shift around the arctic waters when winter comes. If this happens, a beluga can drown because there isn't anywhere to surface for air. Even if it manages to surface for air through a small hole in the ice, it could be killed by a polar bear.

## Try This!

Do you know what the freezing temperature of water is? It's 32 degrees (using the Fahrenheit temperature scale, which is generally used in the U.S.). This means that water freezes when it is 32 degrees or colder. Well, would you believe that the waters in which you can find beluga whales are sometimes below 32 degrees (especially in winter), but they are not frozen? How can this be? Let's find out! You will need two cups, a tablespoon, salt, and a freezer. Fill both cups with water. Add 2 tablespoons of salt to one of the cups and stir. Place both cups in the freezer. Check on them in an hour. Keep checking on them every hour to see what happens. You should see that one of the cups of water either did not freeze or took longer to freeze than the other. Which one was it? Can you figure out why? *See the "Answers to the Narrative Questions" at the back of the book to learn the answer.*

# Narwhals

Do you remember that I mentioned unicorns before? Well, it is time to study the "unicorns of the sea." A type of whale called the **narwhal** (nar' wall) actually has a single, long horn that looks a lot like the horns that have been drawn on the mythical animals called unicorns. In fact, when people found the horn of a dead narwhal washed up on shore, they thought that they had found the horn of a unicorn. At one time, narwhal horns were actually sold as "magical" unicorn horns. Of

This photo of a narwhal was taken from above the surface of the ocean. Notice how long the tusk is.

course, there is no such thing as a unicorn, but the narwhal, which looks remarkably like a gray beluga whale with a horn, is another of God's special creatures of the sea.

The narwhal's horn, called a **tusk**, doesn't grow out of the narwhal's head; it actually comes right out of its mouth. It's an extra long tooth! In fact, some narwhals have two long teeth, making them double-tusked narwhals. Usually, however, a narwhal just has one tusk, and it grows out of the left side of its mouth. A narwhal's body can grow to be fifteen feet long, and its tusk can be more than half that length! The tusk is spiraled like a long piece of clay might look if you twisted it over and over again. This may be why drawings of unicorns always showed them with spiraled horns.

For most of history, scientists have not known what a narwhal's tusk is used for. Some thought they used it for fighting, but there was no way to know for sure. Recently, however, scientists have learned that there are special blood vessels and nerves in a narwhal's tusk. This tells us that the tusk may be used to collect information about the narwhal's surroundings. What kind of information does it collect? How does it collect the information and how is it used? Scientists are still trying to figure out these things. Maybe one day you will help scientists determine exactly what a narwhal does with its tusk!

# Sperm Whales

Sperm whales are the biggest of the toothed whales, able to grow up to sixty feet long. Even a newborn is almost 15 feet long the day it is born! A sperm whale's teeth can be up to eleven inches long! Now, *that's* a toothed whale. Do you remember that baleen whales are often called great whales because they tend to be so big? Well, because sperm whales are so big, they are also called great whales. Sperm whales are the only toothed whales that are called great whales.

Notice that this mother sperm whale and her calf have no dorsal fin and that their heads are longer than those of the other toothed whales you have seen.

The head of a sperm whale is very long, making up a large part of its body. Try to imagine if your head made up more of your body. That would look strange, wouldn't it? The huge head contains an enormous **spermaceti** (spur' muh see' tee) organ. Scientists aren't sure exactly what this organ does, but some think that it is used to focus or reflect sound. Others think it is used to make the head heavier so that the sperm whale can dive more easily. The oil whalers could get from a sperm whale's spermaceti organ was found to be far better than what they got from whale blubber. So the sperm whales were the most hunted whales back in the days of whaling. Whalers could tell a sperm whale apart from other whales because its blowhole is positioned so that water is directed towards the front when it blows out. When a sperm whale surfaces, then, its spout does not go straight up like the spouts of most whales. Also, a sperm whale does not have a dorsal fin on its back like many other whales.

Before diving down into the deep, the sperm whale loves to lobtail with an enormous splash, then dive far below the surface of the water. Can you imagine being a whaler harpooning a sperm whale just before it dives more than a mile below the surface of the water? I'm sure a few ships were overturned in those days. Furthermore, a sperm whale can stay down under the water for more than an hour! Sperm whales are the deepest divers of all marine mammals.

Sperm whales are generally black. A white sperm whale may occasionally be born, but if so, it is usually called an **albino** (al by' no). This is the kind of animal described in the book *Moby Dick*.

Sperm whales consume huge amounts of food each day, including squid. Unlike many other whales, sperm whales don't head to the polar regions each year to find food. Instead, they can be found traveling in small pods in warmer waters year round. These pods are made up of either all females and their young or all males. All-male pods are called **bachelor pods**. Even though it was fiercely hunted in the past, the sperm whale is one of the most abundant whales in the ocean.

*Explain what you have learned about toothed whales.*

# Baleen Whales

Now that you have learned about many of the toothed whales, it is time to learn about some of the baleen whales that God made.  Baleen whales are the largest and most majestic creatures in the sea. They swim through the open ocean on their annual migrations, tenderly care for their young, make amazing splashes, and dive to depths unheard of by most other animals.

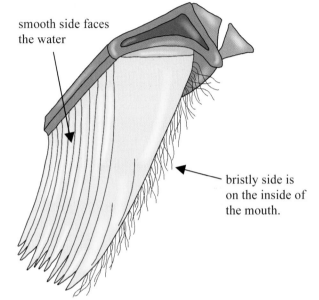

smooth side faces the water

bristly side is on the inside of the mouth.

Each baleen plate is smooth on the outside and bristled on the inside.

You may remember that a baleen whale doesn't have teeth.  Instead, it has baleen, which look like a curtain of long plates hanging from the top of the whale's mouth.  In a large baleen whale, as many as 300 or more plates hang down from each side of the mouth.  These plates might be 12 feet long, and a foot or more wide. Looking at them from the outside, they look like straight knives hanging down, but from the inside, they are bristly, like an enormous toothbrush.

A baleen whale uses its baleen to feed.  It will suck water into its mouth through the baleen.  This filters out the larger creatures in the water, allowing only smaller plankton and other creatures into the mouth. Then, the tongue is used to push the water back out.  Since the *inside* of the baleen is bristly, the small creatures that were let inside the mouth are caught in the bristles, and they are swallowed.

Why does a baleen whale pull in water through its baleen?  After all, if it just opened its mouth and let everything inside, it would have bigger morsels of food, wouldn't it?  Well, it turns out that these enormous, beautiful whales have small throats.  Because a baleen whale can't chew its food (remember, it doesn't have teeth), it can only swallow creatures that fit down its small throat.  Even the huge blue whale has a throat that is not much bigger than a grapefruit.  That's a tiny throat for such a huge animal!  The outside of the baleen, then, acts as a filter, allowing in only the animals that are small enough for the whale to swallow.

# Blue Whales

Though we think of dinosaurs as giant creatures, truly the largest animal that has ever lived on the earth is still alive today, roaming the vast oceans on routes largely unknown by man.  Do you remember which animal that is?  It's the blue whale that I talked about at the beginning of this lesson! It can reach lengths of more than 100 feet, and its head alone can be up to 25 feet long!

Even though we know some things about the blue whale, there is still a lot that we don't know. You see, we have no idea where it goes to give birth to its young. We don't know the paths it takes when it migrates. Isn't it amazing that the largest creature in the world is hard to find, hard to track, and difficult to follow? That just goes to show you how enormous the ocean really is. Even the largest animal on earth can hide from us in the ocean! Interestingly enough, when we find a blue whale, it is either alone or it is a mother with her calf. Because of this, scientists think that blue whales don't form pods. Instead, they seem to prefer to be alone.

This wonderful, mysterious whale can dive deep into the ocean, and when it surfaces, its spout can reach up to 30 feet into the air. Slate blue in color, it has striped grooves under its chin. These grooves stretch out when the whale sucks in enormous amounts of plankton-filled water.

It's simply astonishing that the largest animal in the world feeds mostly on very tiny creatures. Its favorite food is called **krill**. These inch-long, shrimp-like zooplankton swim in swarms, and blue whales feed heavily on them. In the Antarctic summer, these krill are so plentiful that they turn the waters orange with their teeming presence. A blue whale can eat *four tons* of these creatures every single day!

When summer is ending and the blue whale has eaten so much that its blubber supply has increased a lot, it makes its long migration from the polar waters where it feeds to the waters that are its breeding ground. Where are these waters? Scientists are not sure. Wherever these waters are, the blue whale will probably not eat when it arrives there. Instead, it will probably live off the blubber it built up during its feeding frenzy in the waters of Antarctica. Despite the fact that we don't know where it has its calf, we do know that the calf is about 25 feet long the day it is born. It will be about 40 feet long by the next winter!

# Humpback Whales

Can you see the wavy scallops on this breaching whale's fin? They tell you that this is a humpback whale.

One of the most often seen and actively studied whales is the **humpback whale**. It's easy to identify with the warty looking bumps found on the top of its head and its twelve-foot long flippers that have a wavy, scalloped pattern along the edges. It is usually dark on top with a little dorsal fin near the rear of its back. Its underside is white.

As I told you before, humpback whales are best known for the songs that the males sing. They are also known for their amazingly acrobatic ability to breach.  They leap completely out of the water and land with a giant splash.  In addition, they like to go underwater and flap their tails on the surface, spanking the water to make splashes for onlookers.  As you may remember, that's called lobtailing.

# Gray Whales

**Gray whales** love to swim in groups and play in the crashing waves near the shoreline.  They are extremely social with one another and are very intelligent.  Because they are friendly to humans and congregate in huge numbers just off the coast of California in the winter, they are a favorite of whale watchers

As baleen whales go, gray whales are kind of small, growing to be only about 45 feet long.  They are easy to spot with their gray mottled color, which is actually more

This gray whale's blowhole is surrounded by barnacles and other creatures that have hitched a ride.

charcoal black than it is gray.  With all the barnacles and whale lice on it, however, a gray whale does look gray.  Don't feel sorry for the creature because it has lice and barnacles on its skin.  As far as scientists can tell, the lice and barnacles do not harm the whale.  It is possible, in fact, that the lice *help* the whale by feeding off of dead skin, which the whale needs to get rid of.  Many types of whales have lice and/or barnacles that live on them.

Are you right handed or left handed?  Did you know that whales prefer one fin over the other, just like you prefer one hand over the other?  You can actually tell if the gray whale you are seeing is right handed, I mean finned, or left finned.  You only need to notice which side has fewer barnacles.  This is because the whale likes to dive down to the ocean floor to scoop up huge amounts of sand from the bottom, filtering out small creatures that live in it.  When the whale does this, many of the barnacles on the side that rubbed along the bottom are scraped off the whale.  Whichever side has the least barnacles, then, is the side the whale prefers to use when it digs up sand.

Atlantic gray whales were hunted to extinction; so the Pacific gray whale is the only type of gray whale alive today.  Actually, the gray whale was much feared by whalers, and only the very bravest would hunt it.  In fact, when the whalers heard that they were trailing a gray whale, they trembled with fear and wondered if this might be their last whaling adventure, or their last day on earth!  You see, a mother gray whale protects her calf so fiercely that it would actually attack whalers and overturn their boats.  As a result, whalers often referred to the gray whale as the **devilfish**.  Of course, a whale is not a fish.  It is a mammal.

As their name suggests, Pacific gray whales are found in the Pacific Ocean.  They travel from their summer feeding grounds in Alaska down to Mexico for the winter, where they bear their calves.  Since they travel along the California coast, there are stations set up along the coast where you can use telescopes to watch them migrating past.  Of course, you could take a tour to Baja in the winter and enjoy them up close.

# Right Whales

Do you remember that whalers preferred a particular kind of whale?  Which whale was it?  Right – the sperm whale.  Well, in addition to the sperm whale, there was another whale that they liked.  They called it the **right whale**, because it was the right kind of whale to catch.  Even today, we still call it by the same name.

Whalers loved this whale because it has rich stores of blubber and a long baleen.  A right whale's baleen is nine feet long!   In addition, right whales were easy to catch; these 100,000-pound whales were so docile, they didn't fuss or fight when harpooned.  This made them an easy catch for whalers.

Notice the unique mouth shape of this southern right whale.

This whale can be distinguished from others by its mouth, which is in an upside down "U."  It looks like it's frowning!  Look at the picture on the left.  Do you see the white growths on the whale's skin?  Those are rough patches of skin that form wherever the whale has hair.  They are infested with whale lice, which gives them their white color.  These patches are usually found behind the blowhole, on the chin, above the eyes, and on the lower and upper jaw.

# What Do You Remember?

How does a cetacean move its tail to propel itself through the water?  Which is the most important sense to a whale: smelling, hearing, or seeing?  What must a calf do as soon as it is born?  How does the mother help it do this?  Why must a whale have a blowhole?  Where do most whales spend the summer and winter?  Why?  What is breaching?  What is lobtailing?  What is spyhopping?

What is logging?  Why did whalers want to kill whales?  Which two kinds of whales did whalers really like?  How are toothed whales different from baleen whales?  What kind of whale has a "horn" like a unicorn?  Name a difference between dolphins and porpoises.  What is the largest animal on earth?

# Notebook Activities

In your notebook, write down what you have learned about the two different types of whales, whale migration, and whale behavior.  Then, look through books, magazines, or on the internet for photos of different cetaceans.  Paste copies of them into your notebook, and label each kind of whale underneath its picture.  Review these pages so that you will be able to identify different kinds of cetaceans when you see them on shows, in books, or anywhere else.

**Older Students:** Research newspapers at the library or the internet to find a recent report on a beached whale.  Write a speech about this news article, including information that you have learned about beached whales.

# Ocean Box

Add a baleen whale and a toothed whale to your ocean box today.  Remember, you can use clay to make models of them, or you can cut pictures out of magazines or print them from the internet.  You can place them logging on the surface of the water, or perhaps even lobtailing or spyhopping.  Be as creative as you wish.  You can place part of your whale up above the box and part of it inside, as though it were half in and half out of the water.  Do this by cutting the whale in half after you have made it.  Then, glue one half so that it is on top of the box and the other half so it is underneath the top of the box.  See the ocean box pictured on page 17 to get an idea of what I mean.

# Experiment

Whales are able to communicate with one another because sound travels better through water than it does through air.  We're going to do a sound experiment today to explore how sound travels.  This will enable us to understand how sound travels through certain materials better than others.

When someone or something makes a sound, it actually makes waves: sound waves which our ears feel as vibrations.  In order for us to hear something, then, sound waves must travel from where the sound is made to our ears.  When we speak, for example, our sound waves travel through the air around us.  If they reach another person's ears, that person hears us.  When I whisper to my friend who is many feet away, she simply can't hear me, because the sound waves die out before they reach her ears.  But if there were something that my sound waves could travel through better than they travel through air, my friend could hear my whisper even when she is many feet away.  Let's see if sound travels better through air or through yarn.

**You will need:**

♦ Scientific Speculation Sheet
♦ Someone to help you
♦ Two paper (or Styrofoam) cups
♦ A sharp object that can make a tiny hole in the bottom of each cup (like the tip of a straightened paper clip)
♦ A sharpened pencil
♦ 30 feet of 100% cotton yarn (Synthetics tend to stretch too much.)

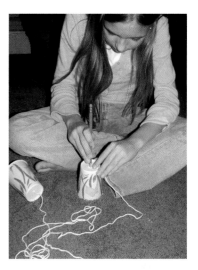

1. On your Scientific Speculation Sheet, record your hypothesis: will sound travel better in air or yarn?
2. Poke a tiny hole in each cup right in the center of the bottom.
3. Use a pencil to push each end of the yarn through the hole in each cup.
4. Tie a triple knot to hold the yarn inside each cup when it is pulled tight.
5. Have your helper hold one cup while you hold the other.
6. Stand in a straight line as far away from the person as you can, pulling the line tight.
7. Have your helper speak softly to you while she holds her cup at her side.  Can you hear her?

8. Have your helper speak with the same volume into her cup while you hold your cup to your ear.  Can you hear your helper now?
9. Try the same thing again, but get just a little closer so the yarn is not tight.  Can you hear your helper now?
10. You can experiment with strings of different lengths if you have them, seeing how long you can make your telephone before the sound waves die out.  You can even experiment with different kinds of string.

   You should have discovered that yarn carries sound waves a longer distance than the air.  In addition, you should have found that this is only true if the string of yarn is stretched tight.  Much like a tight string of yarn, ocean water can carry sound waves a long way, which is one reason cetaceans can communicate with others that are far away.

# Lesson 3
# Seals and Sea Cows

After the 1989 earthquake that shook the coast of California, bringing down buildings, bridges, homes, and lives, something very interesting happened. As if out of nowhere, hundreds of California sea lions suddenly made Pier 39 in San Francisco their home. Even today, they congregate there, barking, waddling, shoving, and jostling for position. Locals and tourists alike enjoy their amazing antics.

**Sea lions** can be found from San Diego to Alaska and in the cooler areas of Australia and New Zealand. These large creatures delight spectators with their dog-like, whiskered faces and barks. Like dogs,

Notice how this sea lion's face looks a bit like a dog's face.

we even call their babies **pups**. Let's take a look at the seals, sea lions, and walruses that pepper the coasts of the earth. After that, we will also look at special kinds of animals called **manatees** (man' uh teez), and **dugongs** (doo' gongz).

# Pinnipeds

**Seals**, sea lions, and **walruses** are mammals that scientists call **pinnipeds** (pih' nih pedz), because they belong to order **Pinnipedia** (pih nih ped' ee uh). The name of the order comes from two Latin words that mean "fin-footed." As you might expect, then, pinnipeds have fins for feet! They also have fins for hands, big bulky bodies, doggish faces, and large eyes. Unlike a cetacean, a pinniped has a nose on its face, and each nostril of the nose closes when the pinniped goes under water.

Like cetaceans, pinnipeds have a thick layer of blubber just under their skin. This blubber keeps them warm in cold waters and keeps them fed during times when food is not easily found. When they can't find food, they can live off the fat in their blubber. At the end of this lesson, you'll do an experiment to understand how blubber works. Although the blubber keeps them warm in cold waters, the waters aren't always cold. To keep cool in warmer waters, pinnipeds will often crawl to shore and lie in wet sand or on other cool, wet surfaces. They also come to shore to rest, get some sun, mate, and give birth. Since pinnipeds have so much blubber, they have large bodies that are more comfortable moving through water than on land. To come ashore, pinnipeds must haul their big bodies out of the water. This requires so much effort that scientists call it a **haul out**.

# Finding Food

Pinnipeds eat a lot. Depending on where they live, they might feast on krill, crabs, clams, fish, or squid. Some, such as the leopard seal, even eat penguins and other sea birds! Pinnipeds usually eat fish or other larger creatures headfirst. They typically use their teeth only to capture prey, not to chew it. If the pinniped isn't holding its prey in the headfirst position, it will often toss its prey into the air so that it flips around and enters the mouth headfirst. This is especially important when pinnipeds eat fish. If a fish is not eaten headfirst, its prickly gills can get caught in the pinniped's throat.

Because their nostrils close when they are under water, pinnipeds mostly hunt by sight and sound (not smell). They have very good eyesight, and many can dive to depths where little or no light exists and still catch prey. Of course, some prey (like the lantern fish) have bioluminescence, which makes them easy to find. Interestingly, a seal called the **crabeater** doesn't eat crabs. In fact, it lives in Antarctica, where there are no crabs for it to eat. Instead, it eats mostly krill, as well as a few small fish and squid.

# Family Planning

Pinnipeds form groups called herds. When it is time to mate, they head to their breeding grounds, which are often called **rookeries** (rook' uh reez). In some rookeries there is only one male and many, many females. Male pinnipeds are called **bulls** and are larger than the females, which are called **cows**. If another bull shows up, there's often a fight. The winning bull stays with the cows and the loser slinks off alone to join a "boys only" group. The bull that's left with the cows is the father of all the pups. Each female gives birth to only one pup per year.

Though they feed at sea, pinniped rookeries are on land. To form the rookery, the bull hauls out first, coming to land to establish his territory. During the whole breeding season, while he is defending his territory, he does not leave his post, even when he is hungry. Thus, he doesn't eat the entire time! Needless to say, once the breeding season is over, the bull is exhausted and hungry, so he leaves. He doesn't help raise the pups at all.

This sea lion pup is nursing. Remember, since pinnipeds are mammals, the mothers nurse their young with milk.

Many of the females arrive at the rookery already pregnant; the pups have been growing inside their bodies since they mated with a bull the year before. The pups are born within a few days of the mothers' haul out, each mother having one pup. Soon after the pups are born, the bull will mate with all the females again.  Because of this, female pinnipeds are almost always pregnant.  In fact, if you see a female seal or sea lion, she is probably pregnant with a pup.

Pinniped mothers care for their pups until they are **weaned**.  This means the young no longer need to eat by nursing.  For some pinnipeds (like the harbor seal), this happens as quickly as four to six weeks, while other pinnipeds (like some sea lions) must nurse for more than a year.  Once the pup is weaned, however, the mother leaves it to fend for itself.  Strangely, some pinnipeds will abandon their unweaned pups when food becomes scarce.  Not only is this bad for the pup, it also causes another problem.  You see, some pups appear to be abandoned, but are only being left alone while the mother searches for food.  Many times, well-meaning folks will try to rescue these "abandoned" pups.  When the mother returns, she finds her pup missing or surrounded by people.  As a result, she retreats, leaving the pup.  Sometimes, she will never come back.

# Pinniped Peril

This great white shark is about to eat the cape fur seal in its mouth.

Pinnipeds eat many kinds of sea creatures, but some sea creatures can also eat pinnipeds.  In the cold arctic waters, polar bears and killer whales are always on the lookout for a scrumptious meal of seal or sea lion.  Pinnipeds are also a popular snack for sharks.  Most often, when seal rescuers find injured seals, the injury is from a shark bite.  However, in addition to these predators, seals face many other dangers.

Sometimes a pinniped pup is caught in a bad storm and is washed away from its rookery.  Sometimes, its mother never returns from a fishing journey.  Other times, pinnipeds get tangled in fishing line or hooks get caught in their skin.  During certain years, when warmer waters come close to shore because of changes in the weather, the fish that pinnipeds feed on are not as abundant, because they have moved into the cooler waters far from the coast.  This creates a problem for mother pinnipeds nursing their young.  During these times, pinniped pups don't get enough to eat when they nurse from their undernourished mothers.  Rescuers will find these pups alone, abandoned for weeks by mothers who went to find food in waters far from the rookeries.  The little pups are sick and malnourished.  When they can, rescuers nurse them back to health.  Sometimes these pups end up in amusement parks, well fed and performing for an audience.

*Tell someone what you have learned about pinnipeds so far.*

# True Seals

If you are looking at a pinniped either at an aquarium, at the zoo, or out in the wild, take note of its head. Does it seem to have ears or not? This is actually how you can tell whether it is a true seal, a sea lion, or a fur seal. You see, true seals are in the family called **Phocidae** (foh' sih day), which comes from a Greek word that means, "seal." All true seals lack one important thing: external ear flaps. This means that when you look at them, they look like they don't have ears at all. They can still hear, however, because they have ear *holes*; they just don't have the flaps that stick out

The pinniped on the left is a true seal. Notice its ear hole and lack of an external ear flap. The one on the right is a sea lion. Notice its external ear flap.

over their ear holes. If a pinniped has flaps that stick out over its ear holes, it is a sea lion or a fur seal. If it does not, it is either a true seal or a walrus. Of course, a walrus is pretty easy to spot (as you will soon see), so you probably won't mistake a walrus for a true seal!

cannot use rear flippers on land

small front flipper ending in claws

can use rear flippers on land

long, hairless front flipper

Notice the differences between the front and rear flippers of the seal (top) and the sea lion (bottom).

Though the ear structure is easy to recognize, there are other features that separate true seals from other pinnipeds. The front flippers of true seals, for example, are short, fur-covered, and end in claws. Sea lions and fur seals, on the other hand, have long, hairless front flippers that end in short nails. In addition, true seals cannot use their rear flippers for walking. So, when they haul out on land, true seals have to move about by flopping along on their bellies. They hold their hind flippers up in the air and slowly inch along the land, something like an inchworm. Sea lions and fur seals, on the other hand, can pull their rear flippers underneath them so that they can actually walk on land.

Now don't feel too sorry for the poor true seal that can't move about very well on land, for what it lacks on land, it more than makes up for in the water! True seals are truly great swimmers. Their flippers are small enough to tuck in tightly next to their bodies, which causes their bodies to be even more torpedo shaped. This reduces water drag, which tends to slow things down in the water. With less drag, they can really speed through the water. Although they usually swim at a leisurely pace of 6 miles per hour, they can reach speeds of up to 18 miles per hour when they want or must! Most seals can stay underwater for 30 minutes before coming up for air, but some seals (like the elephant seal) can stay underwater for up to 80 minutes!

True seals can be found all over the world in every ocean. One kind even lives in a freshwater lake in Russia. Most seals prefer to stay where it is cooler throughout the year. Antarctic ice seals prefer the coldest part of the world, Antarctica, while arctic ice seals prefer the frigid waters in the Arctic Circle. Eighteen different species of true seals can be found today. However, at one time, there were nineteen, but one species (the Caribbean monk seal) has not been seen for so long, it is thought to be extinct.

*Tell someone everything you remember about true seals.*

# Eared Seals

If a pinniped has ear flaps, it is a member of family **Otariidae** (oh tar ee'ih day). The name of this family comes from a Greek word meaning "little ear." As a result, these pinnipeds are sometimes called **eared seals**. There are two basic kinds of pinnipeds in family Otariidae: fur seals and sea lions. Both of these pinnipeds have ear flaps, long front flippers, and rear flippers that can be pulled up underneath the body and used for walking on land.

Now, you may be wondering, "What's the difference between sea lions and fur seals?" The truth is, not much. Nevertheless, there are a few things that do separate the two animals. Sea lions, which are usually larger than fur seals, have more rounded snouts than fur seals. What's a snout, you might ask? It is the part of the face that has the nose. Sea lions also have

The sea lion (left) is bigger than the fur seal (right) and has a more rounded snout. In addition, its flippers are smaller compared to its body than the flippers of the fur seal.

shorter flippers (compared to the body) than fur seals. In addition, a sea lion's fur is rougher and shorter than the fur of a fur seal. In fact, a fur seal's fur feels soft and luxurious.

Because of the pleasant feel of a fur seal's fur, it was made into coats and sold for lots of money, especially in the 1700s and 1800s. It was a big business, and many fur seals were killed for their pelts (furry hides), which almost made some species extinct. In the early 1900s, international treaties were signed to protect the fur seal.

These sea lions can swim very quickly.

Now remember, fur seals and sea lions have longer flippers than true seals. In fact, their flippers can be more than one-fourth the length of their bodies! Because they have such long flippers, they can really move through the water. Sea lions, for example, can reach speeds greater than 25 miles per hour when they swim. That's a lot faster than a true seal can swim. Why do you think sea lions are faster than seals? You're right! It's because their flippers are longer!

Not only are sea lions good swimmers, they are also excellent divers. Like most true seals, they can stay underwater for about 30 minutes at a time. Some sea lions have been observed diving to depths of more than 500 feet!

There are fourteen species of fur seals and sea lions. They live mostly along the Pacific coast in North and South America. They can also be found in Australia and New Zealand and many smaller islands. If you are interested, you can do a research project to learn the fourteen different species of sea lions and fur seals and mark on a map where each of them can be found.

Male sea lions have thicker fur around the face and neck, which sometimes looks like a lion's mane. That's how they got the name "sea lion." Some, like the Steller sea lion, can actually roar! Of course, these pinnipeds aren't really lions at all, but they do have some things in common with lions. Can you think of anything? Well, lions are mammals just like sea lions, so they have fur, breathe air, give birth to live young, and nourish their young with milk. They also both eat meat, even though sea lions eat the meat of fish, squid, and sea birds instead of the meat of land animals. Sea lions have something else in common with the lions you find on land. They are fierce fighters. Especially when they are in their rookeries, male sea lions will ferociously defend their territory from other male sea lions.

*Can you explain the special features of eared seals? In your own words, tell someone the differences between a fur seal and a sea lion.*

Notice that the fur on this male sea lion's neck is thicker and more pronounced than the rest of its fur.

Using what you have learned so far, try to pick out which of the photos below are of fur seals, sea lions, and true seals.  The correct answers are in the back of the book, in the section titled, "Answers to the Narrative Questions."

# Walrus Family

The last and biggest animal in the pinniped group stands alone.  Well, sometimes it waddles, hops, and rolls around alone, too.  Actually, it doesn't really do much alone.  What I mean is that it stands apart from the other pinnipeds, because there's only one species of walrus currently alive in the entire world.

There is only one species of walrus living in the world today.

Walruses are in family **Odobenidae** (oh' doh bin' uh day), and they have things in common with both true seals and eared seals.  Like true seals, they simply have a hole for an ear; they have no ear flaps.  Like eared seals, they can rotate their hind flippers forward to walk on land.  Like both true seals and eared seals, they are excellent swimmers and divers.  However, walruses have several features that neither true seals nor eared seals have.  By looking at the picture above, can you guess what one of those features is?

Besides the fact that their long whiskers make them look like old men, you probably mentioned their **tusks**. And yes, we do call them tusks, but I'm going to tell you a little-known secret about these walrus tusks. First, look at the picture of the walrus. Can you see where the tusks are attached to the walrus? Why, it looks as if the tusks are coming right out of its mouth, doesn't it? We call them tusks, but they are actually teeth! Do you remember a cetacean that had this feature? The narwhal! Walruses have extra long teeth that grow and grow right out of their mouths. The tusks grow for about fifteen years before they reach their full length, which is about 40 inches for the males and 30 inches for the females. Can you get a tape measure and see how long the tusks grow? I think you will be amazed.

With many mammals, only the male grows tusks, but, as I revealed, both the male and female walruses grow tusks because they use them for many purposes. One reason they have tusks is for protection. They can use their tusks to fend off polar bears and killer whales, for example. They also sometimes use them when they eat. When they eat bigger prey, such as seals or small whales, they first tear them apart with their tusks to make them easier to eat.

Another very important use of a walrus's tusks is to help the large animal haul out. Pushing its tusks into the ice, a walrus gets extra help hauling its enormous body out of the sea. This is probably how walruses got their family name Odobenidae, which comes from Greek words that mean "one that walks with teeth." They don't really walk with their teeth, but it can appear that way as they pull themselves out of the water. Their tusks can also be used to cut holes in the ice. These holes are their doorways to the ocean below the ice. Sometimes walruses can be found sleeping with their tusks anchored into the ice. This is probably so that they won't slip around on the ice when jostling occurs between many walruses trying to find a nice place to sleep. Of course, you probably also guessed that the males use their tusks for more unpleasant things like fighting with other males. Indeed, the males use their tusks for showing one another just who's boss.

Like most pinnipeds, walruses travel in herds. Usually, the herd is either all male or all female. When it is time to mate, however, a male will haul out on land and define a territory that will become its rookery. If he sees another male trying to muscle into his territory, he will often hold up his head, showing his tusks to the other male. If the other male doesn't move over, the males will probably fight for dominance. Most likely, there will be bloodshed in the fight. Sadly, walruses can be rather violent in their power struggles; sometimes innocent pups get crushed if they are found near two fighting males.

Look at all the walruses in this rookery!

As with seals, the male that keeps the rookery is the only male that can mate with all the females in the rookery. The calf (that's what a baby walrus is called) grows inside the female for about 15 or 16 months and is usually born on an ice floe. All females in a herd help take care of all the new calves, which nurse (get milk from their mothers) for about two years. When they are weaned, males leave to join an all-male herd, while the weaned females stay with the females that raised them.

Walruses are huge, even compared to the large male sea lions. Fully grown, they can weigh more than 3,000 pounds! The walruses that live in the Atlantic Ocean are a little smaller than those that live in the Pacific Ocean. Both Pacific and Atlantic walruses stay well to the north, in the colder regions of the world. With all their blubber, which may be as much as 6 inches thick, they are perfectly comfortable in icy-cold water.

As huge as these animals are, you might expect that they prefer to eat large fish to maintain their weight. However, these animals prefer the benthic creatures of the sea: worms, shrimp, crabs, and clams, for example. Occasionally, they may eat a fish, seal, or small whale if the benthic pickings are slim. They feed several times a day and can eat more than three thousand clams each day. So how do they find that many benthic creatures? Well, God gave them incredible instincts and special features just for this purpose. They dive down to the bottom of the seafloor and move their noses along the bottom, scouting it like a pig does as it roots in the dirt for food. Their whiskers help them find the benthic animals. Often, while scouting the seafloor, they'll take in mouthfuls of water and blast jets of water at the seafloor, upsetting the sandy bottom, unearthing the benthic creatures that have burrowed beneath. Scientists used to think that walruses used their tusks to dig in the sand to find their food, but we now know that this idea is not right. They use their tusks for the many purposes I mentioned before, but they do not use them to dig for food.

Walruses also change colors depending on how warm they are. They are usually different shades of brown, but as they get warmer, their skin can actually turn pink. This is because as they get warmer, blood rushes to the skin to try to cool it. The extra blood makes the skin look pink. When they are very, very cold, their skin can look almost white. That's because as the skin gets very cold, the animal's body directs blood away from the skin. The lack of blood in the skin makes the skin turn lighter, until it becomes nearly white.

The walrus in the middle of the picture is colder than the others, which is why its skin is lighter in color.

Walruses have an advantage over many other sea mammals when they are in the water. This is because they have air sacs in their necks that they can inflate, allowing them to float as if they were wearing a life preserver. Being mammals, they must breathe air, so they can come up, floating on the surface to rest with their huge air sacs inflated. If a walrus must haul out in order to rest and breathe, it prefers to do it on large floating ice patches instead of land. If a walrus is too close to land, it can find itself confronted by a polar bear, which can feast for quite some time on only one of these huge creatures.

# Manatees and Dugongs

Manatees are slow-moving, docile animals.

**Manatees**, sometimes called **sea cows**, are not pinnipeds. Remember, the animals you have been studying in this lesson have all been a part of order Pinnipedia. Now, however, we are moving on to another order: **Sirenia** (sy ree' nee uh). The word "siren" comes from an ancient Greek myth concerning a man called Hercules. The story goes that aquatic girls called sirens (or mermaids) lived in the sea. These sirens sang very beautiful songs that would put sailors in a trance. The sailors would be hypnotized by these songs, and in trying to find the sirens, they would crash their ships into the rocks and sink, never to be heard from again. Just like the story of Hercules, these stories of sirens were made up, but for a long time people did believe that mermaids were real. Even Christopher Columbus believed that mermaids were real.

When Columbus came to the waters around Haiti, he wrote in his logbook that he had seen three mermaids. He commented that they weren't the beautiful creatures everyone made them out to be. Nowadays, we think that Columbus had actually seen manatees. Perhaps he witnessed them kissing one another, which is something they do when they meet other manatees in the water. One will grab the fins of another manatee it meets, pulling the other manatee close to give it a big smack on the lips.

You may have heard of manatees. They look like gray walruses with no tusks, but are really a totally different kind of animal. They are much gentler than walruses. Even male manatees are calm and quite reclusive; you won't find them swimming or congregating in groups.

We sometimes call them sea cows because, like cows, manatees eat enormous amounts of vegetation. They eat it all day long, chewing and chewing about eight hours every day. Like cows, they are also sweet, harmless creatures. Sea cows have torpedo-shaped bodies with very little hair, giant paddle-shaped tails, and smaller paddle-shaped front flippers, which they use to crawl and paddle along the swampy rivers and oceans where they live and feed.

There are two kinds of sea cows, manatees and **dugongs** (doo' gongs).  Manatees are found in North and South America as well as some parts of Africa, while dugongs are found near Africa, Australia, and Asia.  The main difference between them is the shape of the tail.  A manatee's tail is shaped like a beaver's tail (rounded), while a dugong's tail is shaped like a whale's tail (more triangular).  There are some other differences between manatees and dugongs.  While manatees are often found in freshwater, dugongs are rarely found there.  They spend most if not all of their time in saltwater.  In addition, while neither male nor female manatees have tusks, male dugongs often have tusks.

The tails of dugongs (top) and a manatee (bottom) are quite different.

Manatees have thick pads around their lips with stiff bristles that help them grasp food and move it toward the mouth.  Like all members of its order, a manatee eats sea grass and all manner of aquatic vegetation.  All its teeth are shaped like your molars (the big blunt teeth in the back of your mouth).  The eight hours of chewing each day causes a lot of wear and tear on the teeth.  In fact, its first two pairs of teeth eventually get so worn out that they just fall out every now and then.  However, God compensated for this by giving manatees the ability to continually grow new teeth, just like a shark!  New teeth are always growing in the back of the manatee's mouth.  When the front teeth wear down and fall out, the back teeth move forward to take their place.

Manatees often swim and eat alone.  You may occasionally find a couple together, and while they might be a pair mating, they're more likely to be a mother and child.  That's because when manatees have calves, they nurse them for as long as two years.  Calves stay very close to their mothers, communicating through squeals and calls that travel well under the water.  Though they are mostly solitary, they are very friendly to other sea cows they meet.  Do you remember what a manatee does when it sees another sea cow?  It grabs the other one and yanks it toward itself to give it a big kiss on the face.

Manatees like warm waters.  The Florida manatee, for example, roams north of the Florida coast during the spring, summer, and fall, but in winter, it heads down south where it is warmer.  However, some will find warm waters near power plants that use water as a coolant.  What happens is this: the power plant sucks in cold water from the ocean or an inlet, using the water to cool off the machines being used there. Then, the warmed-up water — still clean, fresh, and unpolluted — flows out of the plant and back into the ocean. The only thing that is changed is the water's temperature. Manatees find these areas where warm water is constantly available and they simply stay there rather than continuing on south. Owners of these power plants have even built little manatee sanctuaries for them, with fresh clean water trickling down from hoses for the manatees to drink, and signs that prohibit boaters from coming into the area. Sometimes bridges are built for tourists to come and watch the manatees from above.

Though they are sea mammals, manatees must have freshwater to survive. They can eliminate excess salt from their diet, but God did not design them to drink saltwater. Thus, manatees must have a constant supply of freshwater. This is often available in small amounts from the plants they eat, but they must usually seek out freshwater to drink. Some Florida residents that live in the inlets and estuaries where manatees dwell are known to leave their water hoses trickling tap water into the ocean. The manatees will come and drink from the hoses every now and then. Wild life experts believe this is actually harmful to manatees, as it makes them dependent on humans and keeps them from venturing to more natural places to find freshwater.

# Manatee Menaces

Manatees are gentle creatures with few defenses. A manatee is slow and therefore cannot really escape dangerous situations. It isn't even able to turn its head to see if anything is coming up behind it! It has to laboriously turn its entire body around just to look behind it. It also has no way to protect itself from predators. It doesn't have sharp teeth or a way to use any part of its body for protection. It doesn't even have a huge amount of blubber to keep it warm in cold regions, which is why it likes warm water. In other words, a manatee is vulnerable, which means it is mostly defenseless against danger.

One of the biggest threats to manatees comes from boats. You see, manatees prefer to stay in shallow waters. Why? Well remember, manatees eat vegetation in the water, like grass. Grass and other vegetation need sunlight in order to grow, and the shallower the water, the more light that gets to the plants that live there. Thus, manatees like shallow water because food is plentiful there. Also, a manatee sleeps at the bottom of its body of water, but every few minutes, it rises to the surface for a breath and then drops back down to the bottom. It is easier to do this in shallow water.

What else do you find in shallow waters? You find a lot of boats. Since there are a lot of boats where manatees stay, and since manatees can't swim very quickly, boats often hit them. Many manatees that forage around the Florida coast have scars all along their backs and tails from the boats that have barreled into them. If a boat is going extremely fast, the force of the accident can kill the manatee. However, because manatees are sturdy with thick, dense rib bones protecting their bodies, boats going more slowly will not necessarily kill them, but can definitely injure them.

States that have large manatee populations have developed laws to protect the manatees. In fact, the entire state of Florida is a manatee sanctuary. If an area is heavily populated with manatees, there will be signs prohibiting boats from entering that area. If manatees are occasionally found in that area, there are signs that warn the boats to go very slowly. Because God has given people dominion over the earth and all the animals, we should protect His magnificent creatures and find ways to keep them safe.

# What Do You Remember?

What is the main difference between a true seal and a sea lion? What are the differences between a fur seal and a sea lion? What is a haul out? What is a rookery? What are some dangers to pinnipeds? How does a walrus differ from other pinnipeds? What does the walrus family name, Odobenidae, mean? What is the main difference between a manatee and dugong? What temperature of water do manatees like? What do manatees do when they meet one another? Why do manatees need to stay in shallow water? Why is this dangerous for them?

# Notebook Activities

Record information about true seals, sea lions, walruses, and manatees in your journal. Make illustrations for each one and write down the interesting things you now know about them. There are pages for this activity in the *Zoology 2 Notebooking Journal.*

**Older Students**: Since manatees are gentle animals, some people actually pet the ones they see in the wild. Write a paper giving your opinion about why people should or should not be allowed to pet wild manatees. A page for this paper is provided in the *Zoology 2 Notebooking Journal.*

# Ocean Box

It's time to add some new creatures to your ocean box. You can make a seal, sea lion, walrus, or manatee — or you can make more than one! You can cut pictures out of magazines or print them from the Internet, but the best thing would be to make a model out of clay. You might even want to make your model look like a seal that is resting on a rock (like the model pictured on the right). Then, you could place it on top of your box so that it looks like the seal is resting on a rock that is sticking above the water. Alternatively, you could make your model so it looks like the animal is swimming through the water.

# Experiment

Do you remember that whales and pinnipeds have a layer of blubber under their skin? How does this help them? Of course, you learned that God provided this blubber to keep them warm in the freezing waters in which they spend most of their lives. The water is so cold that it's actually below 32 degrees in some places, which is below the freezing point of water! If you have worn a winter coat, you know it helps to keep you warm, but can blubber really make it possible for an animal to live in icy water on a cold day? Let's do an experiment to see how well blubber works.

**You will need**:
- A "Scientific Speculation Sheet"
- A big bowl of ice water
- A large container of petroleum jelly or shortening
- Latex or dish gloves (You'll need four for every child doing the experiment.)
- A kitchen timer clock (or a stopwatch)
- Someone to help you

How long do you think you could hold your hand under icy water? How long could you hold your hand under the icy water if you had a layer of blubber? Could you hold it under a few seconds longer? Twice as long? Three times? Let's find out. Make a hypothesis, record it on your "Scientific Speculation Sheet," and then begin your experiment.

1. Put two rubber gloves on one hand.
2. Put one glove on the other hand.
3. On the hand that has only one glove, have someone put a thick layer (about one or two inches) of petroleum jelly all over the glove. Although petroleum jelly is not blubber, it behaves enough like blubber to be used in this experiment.
4. Once the thick layer of petroleum jelly is all over the glove, you and your helper need to put the second glove over that layer of blubber. It might not fit completely. Just do the best you can.
5. Place the hand that does not have the layer of petroleum jelly in the ice water and start the timer at the same time.
6. Use the timer to measure how long you can hold your hand under the ice water before you have to pull it out because it gets too uncomfortable. Record that on your "Scientific Speculation Sheet."
7. Do the same things, but this time use the hand that has the thick layer of petroleum jelly on it.
8. Compare the two times you measured. Now remember, each hand had two gloves, so the protection provided by the *gloves* was the same for each. Now you know why God provided whales and seals with blubber.

# Lesson 4
# Aquatic Herps

We have spent a lot of time discussing sea mammals. Can you remember what makes a mammal a mammal? Well, today you will be introduced to **reptiles** and **amphibians** (am fib' ee uhnz), otherwise known as **herps**. "Herp" comes from the Greek word *herpeton*, which means "creeping, crawling creatures that move about on their bellies." In science, we use the term to talk about reptiles and amphibians. A person who keeps and breeds herps is called a herpetoculturist (hur pih' tuh kul' chur

This olive sea snake is an example of an aquatic herp.

ist) and the hobby is called herpetoculture. The study of these animals is called **herpetology** (hur' pih tahl' uh jee). We won't be studying all herps in this book. We will only study aquatic herps. Aquatic reptiles include sea snakes and sea turtles, and aquatic amphibians include some salamanders, a few toads, and certain frogs.

Are you wondering why a herp is not a mammal? There are many things reptiles and mammals have in common. For example, they both must breathe oxygen from the air, and they both have a backbone, so they are both called **vertebrates**. However, unlike mammals, herps are **ectothermic** (ek toh thur' mik), which means they are cold-blooded. You'll learn more about what that means in a moment. They are also **oviparous** (oh vip' ur us). That means they lay eggs. If an animal is **viviparous** (vye vip' ur us), it means the animal gives birth to live young (like mammals). Both amphibians and reptiles are ectothermic, oviparous vertebrates. Do you understand what that means? That's great! You are really becoming a scientist!

# Ectotherms

So exactly what does it mean to be an ectothermic animal? Well, it means the animal gets its body heat from its surroundings — the air, water, or ground. An ectothermic animal needs to be warm to move around and eat. If it gets too cold, everything in its body will slow down and become sluggish. In cold weather, ectothermic animals like herps can sort of shut down, almost like they are dead. Of course, they aren't really dead. Herps go into a state that is similar to hibernation. It's not called hibernation, though. It's called **brumation** (brew may' shun). Herps brumate, mammals hibernate. Brumation is different from hibernation because the herp is not in a deep sleep. Instead, the herp's body slows down so much that it hardly uses any energy during the whole winter. It's still awake, but very slow and sluggish. This sluggishness causes it to use very little energy. It's sort of in a state of slow motion, which keeps it from needing much food or energy.

Different kinds of herps brumate in different ways. Most freshwater turtles, frogs, and salamanders swim down to the bottom of the pond and burrow into the mud and leaves there. During brumation, their hearts only beat once every few minutes and they quit breathing with their lungs altogether. Even though they don't breathe with their lungs, God designed them to take in small amounts of oxygen through their skin if needed. They can stay like this for months. Some herps, such as snakes and land turtles, burrow into holes underground and brumate through the cold season. It is important to note that some herps don't brumate at all. Sea turtles, for example, don't need to brumate. They stay in warm waters so that they never get too cold.

# Turtle Tales

Sea turtles live their entire lives in the ocean.

Turtles are in order **Testudines** (test uh deen' eez). I'm sure you have seen turtles before, since there are so many different kinds — from tortoises and pond turtles to snapping turtles and sea turtles. Although most turtles can swim and enjoy the water, many are considered land animals. We will discuss them in our next zoology book. In this book, we'll take a look at sea turtles — the only truly aquatic turtles — for they live their entire lives at sea. How long do you think that life is? Strangely enough, we don't really know. No one has actually been able to observe enough of them from birth to death to get an accurate number. However, most scientists think they live about 40 to 60 years, perhaps a little more.

Although sea turtles do spend their entire lives swimming in the ocean, the female does come to land to lay her eggs. I will tell you more about that later.

# Significant Shells

What is the one thing that separates a turtle from other reptiles? Did you say its shell? That's correct! A turtle's shell is made up of an upper dome called the **carapace** (kehr' uh pace) and a lower plate called the **plastron** (plas' trun). Both of these parts are covered with plates (called **scutes**) put together like a jigsaw puzzle. The scutes are made of the same substance as your fingernails, **keratin** (kehr' uh tin). As the turtle grows, its shell grows with it. This can happen when the turtle sheds old scutes and replaces them with new, larger scutes. Other turtles form new layers of keratin around the outer edge of each scute, which look a bit like tree growth rings.  People have said that you can count each of these rings to tell how old the turtle is. This is not always accurate, but it is helpful to scientists.

The carapace and plastron meet at the sides with openings for the head, feet, and tail.  In spite of all the cartoons showing otherwise, you can't remove the turtle from its shell.  This is because its bones (its ribs and backbone) are actually attached to the upper part of the shell.  You couldn't pull the turtle out without leaving its skeleton behind!  Can you imagine going somewhere without your skeleton?  That would be hard!  Now remember, the shell is *covered* with scutes made of keratin, but the shell itself

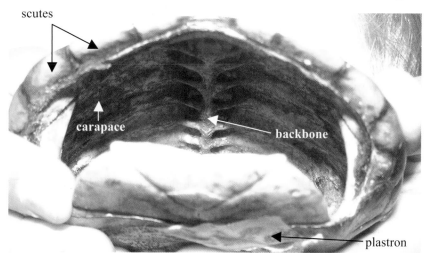

Can you see the backbone and ribs on the carapace of this shell?  It should be clear that there is no way to separate them from the shell.

(carapace and plastron) is made out of bone.  In other words, the shell is actually a part of the turtle's skeleton.

In cartoons, you will see turtles pull their necks into their shell, and even their feet as well.  Actually, a sea turtle cannot escape into its shell.  Although many kinds of land turtles can do this, a sea turtle can only pull its head in a little.  It can't get its head, legs, or tail completely inside its shell.

Sea turtle shells are flatter than those of other turtles.  This gives them a more streamlined shape, which helps them glide through the water, wrestle with ocean currents, and make quick escapes from turtle-eating sea creatures.  Though they can soar through the ocean, flying like a bird under water, they don't do very well trying to get around on land.  Their flippers aren't very good for walking, so the only way they can move about on land is to very slowly drag themselves from one place to another.  No wonder only the female leaves the ocean, and then only when she has to lay eggs!

## Try This!

A female sea turtle uses her front flippers to pull her body hundreds of feet up the beach so she can lay her eggs.  The arm muscles must be very strong to do this, as sea turtles weigh from 75 pounds to more than 1,000 pounds.  That's a lot heavier than you!  Let's see how hard this is.  Lie down on your stomach on the floor, with a lot of space in front of you (go outside in the yard if you need more room).  Put your hands on your shoulders and keep them there.  Now see if you can use your bent arms alone to crawl 20 feet forward.  Do not use your knees to boost you forward.  Don't use your legs to push you forward.  Try to do it with your arms alone.  Was that hard?  Sea turtles make this trip carrying much more weight than you do on arms that are sometimes smaller than yours. Yet, no matter how awkward they are on land, they are like graceful flying birds in the water.

# Give Me Air

While swimming, a sea turtle must come up for air every few minutes to breathe. Its nostrils are located on the top of its snout to

along the ground.

make breathing easier in the water. So, if a sea turtle is swimming near the beach, you are likely to see it at some point, since it doesn't stay under water for long.

When at rest, sea turtles can stay under water for up to five hours without needing to come up for air. Can you guess why they don't need to come up for air as often when they are at rest? It's because while swimming they use up more oxygen than they do while at rest. This is the same for you and me. Did you ever run so hard you began to pant? That was your body's way of getting more oxygen. Just like the sea turtle, when you are at rest, you need less oxygen, too.

# Munching Mouths

beak

Sea turtles have beaks and do not have teeth.

Do you know anyone that doesn't have teeth? When people don't have teeth, they get false teeth. We need teeth because we use them to chew the things we eat. Well, turtles get along quite well with no teeth. So, how do they eat without teeth? God designed them with a strong beak-like mouth. A sea turtle typically has hard ridges along the edges of its jaw and a powerful bite. In fact, the bite is so powerful that it can cut a lobster in half. In order to eat, then, a sea turtle just bites off a piece of its prey and then swallows the piece whole. Sea turtles eat both animals and vegetation; they are omnivores like us.

Many sea turtles begin life eating small sea animals (like jellyfish, shrimp, and fish), and then they switch to a strictly vegetarian diet as they age. The leatherback and the hawksbill turtles, however, love to eat jellyfish, even when they are old.

# Hatching Heroes

Sea turtles lay eggs, which they bury in the sand of a beach. A single mother can lay hundreds of round eggs over a period of several days. Once the eggs are laid and buried, the mother never returns to them. Unlike the animals we studied in the previous three lessons, mother sea turtles do not care for their young. As a result, the baby sea turtles are at risk as soon as the eggs are laid. Millions are killed each year by animals that either eat the eggs or eat the tiny hatchlings when they hatch.

Because so very few sea turtles survive to adulthood, many laws have been passed to protect sea turtles and give them a better chance of survival. If you vacation on the beaches of Florida, you will find signs on every beach that tell you about the laws concerning the sea turtles that nest there during the spring and summer. You see, during the late spring, female sea turtles crawl out on those beaches to lay their eggs. Most scientists think that the mothers actually return to the very same beach upon which *they themselves* hatched!

A mother sea turtle can lay hundreds of eggs in a season.

These female turtles make incredible migrations for thousands of miles across the open ocean, catching currents and passing through regular sea routes so they can return every two or three years to the same beaches to lay their eggs. Scientists aren't sure how sea turtles find their way over that great distance to the same beach on which they were born.

Once a female arrives, she waits until it's dark outside. Then, she slowly drags herself up the beach, using her flippers, which you have already found out is not very easy. She will heave and pull, dragging her burdensome body for hours, until she is finally beyond the high-tide mark. Why does she have to get beyond the high-tide mark? Well, remember that all turtles must breathe air — even baby turtles that are still in the egg. If water were to cover an egg for very long, the developing turtle inside would suffocate. So, the mother needs to make sure there is no chance that her eggs will get covered with water, even during high tide.

Trying to finish this difficult task before daybreak, she first digs a large depression with her front flippers, then crawls in and begins digging a deeper hole with her hind legs. After this laborious chore, she settles in to lay between fifty and one hundred fifty eggs in the hole. Two or three eggs usually drop out at a time. The eggs look a lot like ping-pong balls, but they are soft and leathery. Inside the egg is a developing baby sea turtle.

Once the eggs are laid, the mother turtle covers the hole with sand and returns to the sea. She will most likely do this same activity several times in different locations on the same beach. When all the eggs are laid, she leaves the coast and heads out to the ocean on a long journey that scientists have still not been able to fully understand. She'll return in a year or two to do the same thing again.

Buried in the sand, the turtles inside the eggs are at risk. Predators such as raccoons, foxes, and coyotes like to dig up and eat turtle eggs. Ants may also find them and feast on them. Depending on how warm the sand is, they will hatch in about two months — if they survive.

Wouldn't it be strange if only boys were born in frigid, cold areas of the world and only girls were born in tropical, warm areas? Think about what the world would be like if during the warmer years, most of the babies born were girls and during the colder years, most of the babies born were boys. Yet, would you believe that the temperature of the sand determines whether the hatchling turtles are mostly male or female? Males tend to hatch from eggs buried in cooler sand, and females tend to come from eggs hatched in warmer sand.

# Sand Flight

When they are ready, the babies break out of their shells and begin clawing their way out from under the piles and piles of sand in which they are buried. It might take several days for them to reach the surface, but when all of the baby turtles have hatched, it becomes a collective effort, and the progress is faster. It's a hard job, but they must get out of the sand in order to survive.

These sea turtle hatchlings are trying to make it to the ocean before they get eaten by a predator.

When they get near the surface, they usually wait for the sand to cool, which tells them it is night. At that point, they dig themselves out the rest of the way, and then the little, tiny turtles make a dash for the sea. Now, how on earth do they know which way to go to reach the ocean? Well, scientists have learned that they head toward the brightest horizon they can see. Since the sea is nice and flat, and since the moonlight often reflects off it, the sea is a bright horizon at night. At least that's how it was for thousands of years. Today, however, lights from houses or hotels on the beach can distract the hatchling sea turtles. Many hatchlings are led away from the ocean because they are drawn to these manmade lights, and as a result, they usually end up dying. Thomas Edison may be a hero to us, but he's not to the sea turtle!

The baby sea turtles' journey to the sea is fraught with many dangers. Besides the danger of heading in the wrong direction, sea turtle babies are a delicacy to many shore animals. Sea birds, crabs, foxes, lizards, and other predators will gobble up many of the turtles before they even reach the sea. This is why they wait until night to start their journey. The darkness at least partially hides them from predators. Only a few end up making it to the ocean; some studies say it is only one in a hundred! Now before you feel too bad about this, remember that it is all a part of God's plan. The baby sea turtles are the food that God provides to the predators, and He makes sure that sea turtles do not become extinct by designing the mother sea turtles to lay hundreds of eggs. That way, even though most of her babies will die, some will survive, which keeps the sea turtles in existence.

*Explain what you have learned so far about sea turtles.*

# Eight Turtles of the Sea

There are eight species of sea turtle in the earth's oceans today. The **loggerhead** can often be spotted in the warm waters off the coast of the U.S. It can grow to just over 3 feet in length. This species got its name because its head is large (up to 10 inches wide) and block-like. With powerful jaws, it searches the seafloor for lobsters, crabs, and any tasty animals it might find there. Although it is the most common sea turtle found off the Florida coast, it is considered an endangered species.

Loggerhead turtles are named for their large heads.

Green sea turtles are not really green in color.

Although it might have some green on its carapace, the **green sea turtle** is not green. It gets its name from the fact that its body fat is green. It is bigger than the loggerhead, growing up to 4 feet long and weighing more than 500 pounds! Although they are an endangered species, green sea turtles can still be found in enormous numbers on the east coast of Florida. You can also find them in other warm waters around the world.

Green sea turtles begin life as an omnivore, eating plants and algae as well as plankton and other small animals. As they get older, however, they become mostly herbivorous, eating only vegetation. In other words, they "decide" to become vegetarians as they get older. In fact, their body fat is green because of all the vegetation they eat as adults. Their jaws have serrated edges (like the edge of a saw), which helps them to rip off pieces of thick vegetation so that they can swallow the pieces whole.

The **leatherback** is the champion of sea turtles: growing the largest, diving the deepest, and traveling the farthest of all sea turtles. A leatherback can weigh up to 2,000 pounds and can reach lengths of greater than 9 feet!

Why do we call it a leatherback? Can you take a guess? Well, instead of a shell, the leatherback grows a tough, leathery carapace on its back. Seven thin ridges run down the length of the carapace, which is usually black with white spots.

This mother leatherback is trying to get back out to sea.

Amazingly, this creature was specially designed by God to withstand cold water temperatures that would send most reptiles into brumation. Do you remember what brumation is? It is when a herp rests so deeply that it uses very little energy. The leatherback can swim in waters that are about 40 degrees, just above the freezing point of water!

Unlike the loggerhead, with its shell-crushing jaws, the giant leatherback has delicate, scissor-like jaws. These jaws are designed to eat only the softest of animals. What do you think is their favorite animal to eat? Jellyfish! They fill up on them, eating twice their weight in jellyfish each day. Giant jellyfish, which you'll learn about in a later lesson, are found far out at sea. That's where you'll also find leatherbacks.

One threat leatherbacks face is plastic grocery bags floating in the water. You see, they can look like jellyfish to a leatherback, and as a result, the leatherback eats them. People have found dead leatherbacks with these bags in their mouths, having choked on what they thought was a good meal.

## Try This!

Go outside and use a tape measure and chalk to draw a leatherback turtle on a flat surface, like a driveway. You could also use masking tape to make an outline on the carpet. Make it 9 feet long and 6 feet wide. You can use the picture above as a guide for what it should look like. How would you like to run into a leatherback while swimming in the ocean? Well, it might seem frightening, but leatherbacks are not aggressive creatures. In fact, you might enjoy having them nearby, for they will devour all the jellyfish in the area! That would be good for you, because a jellyfish's sting can cause you a lot of pain.

Weighing up to 200 pounds and measuring about 3 feet long, **Australian flatbacks** are thought of as "mid-sized" sea turtles. Preferring warm coral reefs, they feast on sea cucumbers, jellyfish, and other invertebrates. They can only be found in one spot on earth. Can you guess where that might be? Look at the name. Right! They can only be found off the coast of Australia. There's something else you can learn about this turtle by just looking at its name. Can you guess what it is? The Australian flatback has a flat carapace, so it has a very flat back.

Smaller than the Australian flatback, the **hawksbill** sea turtle is usually just over 2 feet long, and it generally weighs about 150 pounds. It has the most pointed beak among the sea turtles, which is how it gets its name. The hawksbill gets food from inside nooks and crannies in coral reefs, eating sponges, squid, shrimp, and other invertebrates. Can you think why the hawksbill has such a pointed beak? It helps the turtle get food out of the tiny nooks and crannies that it searches.

The hawksbill turtle has a pointed beak.

Though they are occasionally seen in American waters, hawksbill turtles almost always nest in the warmer climates that run along the equator. Like other sea turtles, the hawksbill is an endangered species, hunted throughout the world for its beautiful shell. People use the shell to make jewelry and other products.

The smallest turtles of all, the **ridley turtles**, are usually about 2 feet long, and they generally weigh less than a hundred pounds. There are two species: the **olive ridley** turtles and the **Kemp's ridley** turtles. The olive ridley turtles are olive green and are typically found in the warm waters around Costa Rica. The Kemp's ridley is olive green to grayish green and is considered the most endangered of all sea turtles.

Ridley sea turtles, like this olive ridley, are the smallest of the sea turtles.

Most Kemp's ridley turtles nest on a small strip of beach in Rancho Nuevo, Mexico. During the highest tide, all at once, they arrive on the beach. The locals call the arrival of the Kemp's ridleys *los arribadas*, which means "the arrival." Once they arrive, they all lay their eggs on the same strip of beach. After three months, the newborn turtles hatch and make a mad dash to the sea. Birds and many other creatures seem to remember when this happens every year and eagerly await the event. Once the baby turtles finish struggling and emerge from their holes, a feeding frenzy follows, and thousands of turtles become meals for hundreds of animals on the beach.

The Kemp's ridley turtles are considered endangered because in 1942, someone filmed forty-two thousand Kemp's ridleys nesting at Rancho Nuevo. Filming again fifty-three years later (in 1995) showed a measly 1,429 Kemp's ridleys nested in the same spot. There are believed to be fewer than two thousand Kemp's ridley nesting females left in the world. Will they still be around when you are grown up?

The **black turtle** is usually seen along the coasts of California and Mexico, but it can be found in other places. Its shell is very dark (almost black) in color, which is how it got its name. An adult black turtle can weigh more than 200 pounds and can grow to lengths of more than 3 feet.

*Explain in your own words all that you can remember about the eight species of sea turtle.*

# Sea Snakes

All snakes are good swimmers, but only true **sea snakes** live permanently in the ocean. Not only do they live there, they eat there and even have their young right in the ocean. True sea snakes can't even survive on land. If they accidentally get forced onto land by ocean storms or powerful currents, they are absolutely helpless, unable to move. Why? Because land snakes have special scales called scutes on their bellies, which give them traction to grip and slither across the ground. Sea snakes have no scutes, so if a sea snake ends up on the beach, it is unable to slither back out to sea.

Notice the paddle shape of this sea snake's tail.

Many snakes that like to spend time in the water head toward the land to lay their eggs. However, true sea snakes are the only marine reptiles that don't come to land to have their young. Instead of laying eggs somewhere, the eggs are stored inside the mother, in a compartment called the **oviduct** (oh' vih dukt), until they hatch. When they hatch, they simply slither out of their mother's pouch. So, here's a question for you. You remember that animals that give birth to living young are called viviparous, and animals that lay eggs are called oviparous; so what is a snake called when she lays eggs inside herself and then gives birth to living young? The answer is, **ovoviviparous** (oh voh vye vip' ur us)! Do you see the two words combined in this word?

So how would you know if you were seeing a sea snake or a land snake that just happens to be swimming? Well, a sea snake has one easily recognizable design feature: a paddle-shaped tail. The tail needs to be paddle-shaped so the snake can swim well. There are other ways to tell sea snakes from land snakes. Sea snakes, for example, also have specially designed nostrils that can close up when they go underwater and open when they come up for air, much like the blowhole of a whale. We say they have **valved** nostrils. As you may have guessed, the nostrils are found high up on the head so that the entire snake can stay submerged in water when it comes up to breathe.

The sea is a rather salty place, and when a sea snake drinks or eats animals in the sea, it can get a lot of salt in its body.  You and I would die out at sea unless we had freshwater to drink.  God created sea snakes to survive on saltwater, however.  He gave each sea snake specialized glands under its tongue.  These glands get rid of excess salt.

As a sea snake forages for food, sometimes it must dive deep down to the bottom of the ocean.  In order to stay under the water, it needs to be able to hold its breath for a long time.  As a result, God designed the sea snake with a lung that is almost as long as its whole body.  If your lungs were as long as your whole body, you could hold your breath for a long time.  Sea snakes can also absorb oxygen from the water through their skin.  As a result, a sea snake can take one breath and stay under water for two whole hours!

# Positively Poisonous

Interestingly, another common design feature of most sea snakes is tremendously poisonous venom.  Of the more than fifty species of sea snakes, almost all are extremely and dangerously venomous.  Though the venom of many snakes takes a while to begin working, sea snake venom is fast-acting, paralyzing its victim almost instantly.  One bite, and the victim quickly finds itself unable to move.  This helps sea snakes catch the slippery prey that they love to eat.  All they have to do is bite their prey, and it cannot escape.

Like almost all sea snakes, this banded sea snake is poisonous.

The fact is, sea snake venom is known to be the most poisonous venom of all snakes.  The most feared sea snake is the beaked sea snake found in the South Pacific.  It has fangs that are so long they can pierce through a thick wetsuit.  One drop of its venom can kill five adult men, and it has enough of this powerful venom that if it was all released, it could kill fifty-three grown men.

Even though their venom makes sea snakes very frightening to people, we are actually much more frightening to them.  You see, most sea snakes are docile, shy creatures; they swim away from contact with humans.  They rarely bite unless provoked.  If one does bite in self-defense, it usually doesn't release venom, because venom is saved for catching prey.  Fishermen, pulling up their fishing nets with a sea snake caught inside, are the most likely people to see a sea snake.  However, in Australia, they are often found washed up on the beaches after violent storms.  Of course, as I mentioned before, they are stranded when they are washed up on land and end up dying.

If you ever see a sea snake, just leave it alone, even if it's dead. A dead sea snake might be just as dangerous as a living one. "How can that be?" you might ask. Believe or not, a sea snake can bite you and release its venom up to an hour after it is totally dead; even if its head has been cut off! This is because the snake's bite reflex remains active for quite some time after it has died.

If a snake bit you, would you rather suffocate or die from a lack of blood flow? Well, you probably wouldn't ever have to worry about either because we have medicine that stops the snake venom from working. It's called **antivenin** (an' tee ven' in). However, if you couldn't get the antivenin in time, those would be your options. You see, snake venom is either a **neurotoxin** (noor oh tahk' sin) or a **hemotoxin** (hee muh tahk' sin). Neurotoxins attack the nervous system, causing muscles, including those that help us breathe, to stop working. As a result, the victim usually dies of suffocation because it cannot breathe. Hemotoxins, on the other hand, go into the blood system and destroy the ability of the blood to travel through the body. That, of course, will also result in death. The venom of some land-dwelling snakes is made of neurotoxins, while the venom of other land-dwelling snakes is made of hemotoxins. All sea snake venom, however, is made of neurotoxins.

While snake venom may seem like the worst thing in God's creation, it isn't. Believe it or not, sea snake venom (and other snake venom) can actually be used to treat diseases. Scientists are finding that neurotoxin venom, when processed properly, may prevent Alzheimer's disease and strokes. In addition, some hemotoxins may cure heart disease. Another thing scientists use snake venom for is to create antivenin medicines, which save you when you are bitten by…a snake!

## Spotting Sea Snakes

This sea snake is looking for food in a hole.

So where on earth would you find a sea snake? You may be wondering this and hoping that they aren't swimming near the beaches you visit. Well, if you live in North America, you're safe. No sea snakes can be found in the Atlantic Ocean or on the cold parts of the Pacific coast. Because they are ectothermic, you won't find these creatures anywhere it gets cold. They live in warm ocean waters, from the Persian Gulf across through Southeast Asia and down to the warm waters off the coast of northern Australia.

Although most sea snakes live close to shore, one species (the yellow-bellied sea snake) lives far from the land in the open ocean. It can be found thousands of miles from land in both the Indian and Pacific Oceans.

Sea snakes come in a lot of different colors; however, many of them are banded. This means the snake has rings running around its body. This makes sea snakes easy to see, so they are not very well camouflaged. Although they are cunning predators, there are animals that eat them. Predators of sea snakes are sea eagles, sharks, large fish (like barracudas), large eels, and crocodiles.

*Tell someone what you have learned about sea snakes.*

# Reptiles versus Amphibians

Do you know the differences between a reptile and an amphibian? Before I tell you, see if you can think of any differences yourself. Think of all that you know about turtles, snakes, lizards, and frogs.

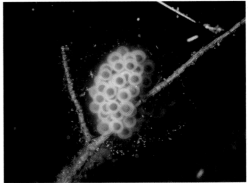

There are many differences between reptiles and amphibians. Two important ones are their eggs and how their young develop. Jelly-like amphibian eggs have to remain moist and are usually laid in water, although some amphibians make juices that surround the eggs. Most, however, deposit their eggs in some pond or other quiet body of water. They are especially easy to find in May or June. Frog eggs are attached to one another in clumps, while toads make strings of eggs that are all attached to one another.

Amphibian eggs must stay moist, and they do not have leathery shells like reptile eggs have.

This is not a fish. It is a baby amphibian, called a tadpole. Note the tiny back legs that are beginning to develop.

Newly hatched amphibians look nothing like adult amphibians. Hatchlings (called **tadpoles**) live completely underwater, breathing oxygen from the water with gills. They go through a series of changes over several months, looking more and more like adults with each change. Eventually, they become adults. Most adult amphibians enjoy the water, but because they can breathe oxygen through the air, they don't live in the water. However, there are some amphibians, which are known as **aquatic amphibians**, that do stay in the water. Some even continue to breathe with gills. We will be studying aquatic amphibians later in this lesson.

The eggs of reptiles are not like the soft, jelly-like eggs of amphibians, but they are also not hard like a chicken's egg. Reptile eggs are soft and have a leathery covering. Inside the reptile's egg is a thick substance that protects the embryo, and there is a yolk to give it food while it grows. Unlike an amphibian, when a reptile hatches from its egg, it is a miniature of the adult. It may change colors as it grows, but a baby snake looks like a tiny snake, a baby lizard looks like a tiny lizard, and a baby turtle looks like a tiny turtle. Most reptiles lay their eggs on dry land, even the entirely aquatic reptiles.

Sea snakes are an exception to this general rule. As you already learned, they nestle their eggs in their own bodies until they are ready to hatch.

Another major difference between reptiles and amphibians is that reptiles have dry scales on their skin, made out of keratin, the same material in your fingernails. The skin cells can clump together in thicker parts making what appear to be horns or spikes. But in truth, it's all just keratin scales; dead skin cells that are really tough. A reptile's skin is dry and can withstand very dry climates, like the desert.

Amphibians don't have scales; instead, they have soft skin that must stay moist or the entire creature will dry out. In fact, if the weather gets too dry, amphibians go into a kind of hibernation called **estivation** (es' tuh vay' shun), in order to survive until it gets wet again.

God created amphibians with a unique kind of skin, for it doesn't just cover the body, it also breathes and drinks water! It's true! Amphibians don't usually swallow water like we do; instead, they absorb most of the moisture they need right through their skin. Not only that, they are able to take in extra oxygen through their skin. Their lungs don't have to do all the breathing for them, but their skin must be moist in order for them to breathe.

## Try This!

Oxygen is an element we need to breathe, and it exists in most bodies of water as well as in the air. To understand how a frog's skin can absorb oxygen from the water, try this experiment: Without spilling, carefully pour a teaspoon of vanilla into a balloon and tie the end so it is tightly sealed. Find a box in which to place your balloon. Smell the box and notice the odor before putting in the balloon. Now place the balloon in the box and close it up. In 30 minutes, remove the lid and smell the box again. Did any of the vanilla get out of the balloon? How can you tell? Even though it was sealed up, some of the vanilla escaped the balloon through a process called **diffusion** (dih fyoo' shun). Oxygen can get into the skin of an amphibian the same way, allowing the amphibian to use its skin to breathe.

# Frog or Toad?

A frog's skin is smooth and moist.

Have you ever caught a frog? What did its skin feel like? If you've never held a frog, it's okay. From looking at the picture on the right, what do you think its skin might feel like? It doesn't look rough or scaly like a snake, does it? No. Frogs usually have smooth skin, smoother than a toad's skin. A frog's skin can feel almost slimy, because it produces mucus that helps keep its skin moist. Frogs are more dependent on wet skin than toads are, so they must always be near a source of water.  If they can't find water, they dry up.  Toads, on the other hand, have tougher skin that doesn't dry out as fast, so they can live farther from the water for longer than frogs can.  Toads usually have bumps all over their skin.

As you already know, amphibians begin life in water, breathing with gills.  As they transform into adults, they usually begin breathing with lungs and lose their gills.  Aquatic amphibians are different.  Many of them keep their gills their entire lives.  Most are found only in freshwater ponds, streams, and rivers.  Let's look at some of the aquatic amphibians in God's amazing creation.

# Aquatic Toads

There are three basic kinds of aquatic amphibians: **aquatic toads**, **aquatic frogs**, and **aquatic salamanders**.  All frogs and toads belong to order **Anura** (uh nur' uh) and are called **anurans**.  The name of this order comes from Greek words that mean "without tail."  In other words, frogs and toads are amphibians without tails.  Let's look at one of God's aquatic toads.

The pipa pipa is so flat it looks like a leaf when it floats in the water.

The **Surinam** (sur' ih nam) **toad**, known as the **pipa pipa**, is an odd little aquatic toad that lives in South America.  It is almost perfectly flat, looking a lot like a dead leaf when it's floating in the water.  It prefers murky, unmoving water where it can be camouflaged.  God designed its wide mouth so that, when it opens underwater, it literally vacuums in any animal that happens to be swimming by it.  This toad also has ultrasensitive feet that can catch fish or other creatures in the water.  It feels around the water and grabs food, sucking it into its mouth.

Laying eggs is a strange process for the pipa pipa.  You see, once the female lays her eggs, the male toad takes the eggs and places them on her back.  "So?" you may be saying.  Well, here's the really interesting part: the eggs begin to embed themselves into the female's skin and actually become a part of her back!  A skin membrane grows over the eggs.  Even more strange, when the tadpoles hatch from the eggs, they don't swim off.  They stay right there, embedded in their mother's back.  Then, when they have gone through the entire tadpole phase and have developed into fully mature toads, they break out of her skin and swim away as tiny adults!

# Aquatic Frogs

There are several kinds of aquatic frogs.  The most popular are the African frogs, because they are often kept as pets.  Fully aquatic, they are just as easy to keep as goldfish.  **African clawed frogs** and **African dwarf frogs** are both aquarium frogs.  Unlike many aquatic amphibians, these frogs actually breathe air with lungs, but like sea snakes, they never ever leave the water.  The African dwarf frog is quite an aggressive little frog and will tend to eat any animal that is smaller than itself.

If you want to learn more about aquatic frogs, there are many science supply houses that sell aquatic frogs you can raise from tadpoles.  In fact, the experiment at the end of this lesson involves getting and raising aquatic frogs.

# Aquatic Salamanders

Salamanders belong to order **Caudata** (caw dah' tuh).  *Cauda* is a Latin word that means "tail," so salamanders are the amphibians that have tails.  Because they are secretive creatures of the night, salamanders are among the least studied of all animal groups.  At the same time, they are considered one of the most plentiful animals in the world.  They love dark, wet places and can live in many different environments, from rain puddles to large rivers, from cold streams to warm ponds.

What kind of animal does a salamander look like?  It looks like a lizard.  Most people think of them as lizards, and some even call them "spring lizards."  Lizards, however, are reptiles, and you know that amphibians aren't reptiles, don't you?  Since salamanders are amphibians, they are *not* lizards, even though they may look similar.

Although most salamanders are just a few inches long, some in the United States grow to be 3 feet long.  Like most amphibians, most salamanders have gills when they first hatch, but those gills disappear as they change into adults, being replaced by lungs.  As babies, then, salamanders use their gills to breathe oxygen under water.  As adults, however, most cannot breathe under water and must breathe the air above the water's surface using their lungs.

gills

Waterdogs are aquatic salamanders with external, feathery gills.

There are some kinds of salamanders, however, that never lose their gills.  **Mudpuppies** and **waterdogs**, for example, have gills their entire lives.  These gills are actually on the outside of their bodies and look like feathery spikes.  They are red because they contain a lot of blood.  Aquatic salamanders that live in warm waters grow larger gills than those that live in cooler waters.  This is because there is less oxygen in warmer waters as compared to cooler waters.

Salamanders eat insects, worms, and other small invertebrates.  Like the pipa pipa, aquatic salamanders use suction to capture prey.  Like sea turtles, they usually return to the same place they were born to lay their eggs. Hundreds might migrate up rivers and through streams into the pond where they hatched. There, they will lay their eggs so that their offspring will come back to that same pond to start the process all over again.

Can you imagine capturing a salamander and finding out that it is older than your mother? It could happen. You see, though most salamanders can live eight to twenty years, one kind of large aquatic salamander, called the **hellbender**, lives to be fifty-five years old, at least in zoos!

You can sometimes find aquatic salamanders by carefully turning over rocks on the stream bottom. Most salamanders don't bite, but some large aquatic salamanders can bite very hard. Salamander skin is very sensitive; so always handle salamanders with wet hands. No salamanders have poisonous bites, but some produce toxins that can irritate human skin.

# What Do You Remember?

What is the name we use to mean both amphibians and reptiles? Name some of the differences between mammals and reptiles. How are sea turtles different from land turtles? What is the top part of a turtle's shell called? What term do we use to describe hibernation in herps? What are some of the dangers sea turtles face? What makes a sea snake different from other snakes? What are the two kinds of snake venom? How are reptiles different from amphibians? What is the difference between most amphibians and aquatic amphibians?

# Notebook Activities

In your notebook, write down what you have learned about each of the eight species of sea turtles. Draw an illustration of each. If you prefer, you can use the Internet to print pictures of them, which you can then paste in your notebook. Also, write a story about a reptile meeting an amphibian. Have them talk about their differences. They can comment on the differences in their skin, and they can each talk about how they grew up. If you own the *Zoology 2 Notebooking Journal,* there are pages for these assignments.

**Older Students:** Do some research to learn what is being done to preserve the sea turtle populations. Write down the results of your research. There is a page for your to record your findings in the *Zoology 2 Notebooking Journal.*

# Ocean Box

Make a sea turtle and a sea snake to add to your ocean box. If you are making models, be sure to give the sea snake a flat, paddle-like tail.

# Experiment

To learn more about aquatic frogs, I want you to raise them. This is actually not very hard. You can sometimes find them in a pet store, or you can order them from a science supply house. The course website given in the introduction to this book has links to several places from which you can order aquatic frogs. They are inexpensive, easy, and fun to raise. All you need is a tank that has a lid with air holes and some food. Once again, these things can be bought at a pet store or ordered from a science supply house. Be sure to read the instructions that come with your frogs.

Of course, if you would rather search for wild frogs to raise, you can find eggs in shallow ponds between the months of April and June. You'll find clumps of jelly-like eggs floating near the surface. Bring them home and provide dried shrimp or tadpole food for the babies to eat when they hatch. Now if you find eggs in the wild, they will probably not be from aquatic frogs. As a result, you will need to provide a dry place for the frogs to rest when they become adults, like a rock sticking out of the water. You will also need food for the adults. You can buy frog food from a pet store, or you can feed them live insects. You can also check the course website to get more details on how to care for frogs you find in the wild.

However you acquire them, be sure to watch them as they grow up. You will be amazed at the changes that occur as the tiny tadpoles become more and more frog-like until they mature into adult frogs. You can even do an experiment with the tadpoles once they have hatched, as long as you have more than one.

## Does Temperature Affect Tadpole Development?

You will need:
- ♦ A "Scientific Speculation Sheet"
- ♦ Two tadpoles
- ♦ Two covered tanks (or large glass bowls) that are the same size
- ♦ A lamp with a 50- or 60-watt bulb

1. You are going to watch two tadpoles grow into adult frogs, but one will be in warmer water than the other. Before you begin, make a hypothesis about which tadpole you think will develop more quickly. Will it be the one in the warmer water, or the one in the cooler water? Record your hypothesis on a "Scientific Speculation Sheet."
2. Put one tadpole in one tank and the other tadpole in the other tank. The tanks can be small, but be sure they are covered with lids that have small holes for air.
3. Shine the light on one of the tanks each day for two hours. This will warm the water inside.
4. Make sketches of each tadpole as it grows. Remember to keep everything else (like the amount of food and the water level in the bowl) the same for both tadpoles. The only difference should be that one tadpole's water is warmed by the light.
5. Use the "Scientific Speculation Sheet" to record your experiment. Was your hypothesis correct?

# Lesson 5
# Primeval Reptiles

Have you ever been afraid of monsters?  Maybe when you were very young you believed that there was a monster in your closet or under your bed.  If so, I am sure that your parents told you, "There is no such thing as monsters!" And your parents were absolutely right.  But down through history there have been reports of very large, hideous sea reptiles that could smash boats and gobble up sailors in a single bite.  Sailors called them monsters — sea monsters, that is! Ancient stories and legends are full of reports of these colossal, consuming creatures.

Hans Egede, a Christian missionary to Greenland wrote, "On the 6th of July 1734, when off the south coast of Greenland, a sea-monster appeared to us, whose head, when raised, was on level with our main-top.  Its snout was long and sharp, and it blew water almost like a whale; it has [*sic*] large broad paws; its body was covered with scales; its skin was rough and uneven; in other respects it was as a serpent; and when it dived, its tail, which was raised in the air, appeared to be a whole ship's length from its body." (Strange Science, "Sea Monsters," by Michon Scott, last modified August 10, 2012, http://www.stangescience.net/stsea2.htm.)

For over a thousand years, many people believed the stories of the sailors.  Even mapmakers drew large sea dragons out in the ocean to indicate that sea monsters lived out there.  As the centuries passed, however, there were fewer and fewer reports of sea monsters, and most people just began to think that the sailors made up the stories.  They laughed at the sailors saying, "They've been out at sea too long." Sea serpents were thought of as imaginary monsters.

This drawing shows a sea monster, like those described by sailors, attacking a ship.

Then something happened that made people think again.  In 1810, a brother and sister from England made an amazing discovery! Mary and Joseph Anning were out collecting seashells and fossils along the seashore.  They sold them to help support their family.  While Mary and Joseph looked for shells and fossils, they always kept their eyes open for anything unusual.  One day, Joseph noticed some very odd-looking stones sticking out from the surrounding rocks.  These stones didn't look like the other rocks scattered on the shore.  They looked like old bones, and that's just what they were.  They were the fossilized bones of a huge, swimming reptile, a creature that looked a lot like the sea monsters described by the sailors of old.  Mary and Joseph decided to dig up the bones.

Eleven-year-old Mary was fascinated with the fossil bones and came back often to continue the hard work of digging them up. The news of Mary's "sea monster" spread to London. Soon, scientists became interested. As the scientists examined Mary's "sea monster," they realized that it was the fossilized remains of a creature that had never been seen before. We now call it an **ichthyosaurus** (ik' thee uh sawr' us). You will learn more about ichthyosauruses in a moment. A few years passed, and Mary made another discovery. Along a seashore full of shells she found the fossil remains of a monstrous sea creature we now call **plesiosaurus** (plee' zee uh sawr' us). Mary and her brother Joseph continued to collect and sell fossils all of their lives. Mary became quite famous, and people began to say a rhyme about her. The first line of the rhyme is, "She sells seashells by the seashore." Have you ever heard that tongue twister? Now you know where it came from. Try to say it three times quickly!

With these and other discoveries, people began to reconsider the sailor stories of old. Perhaps these creatures could smash boats and devour full grown men with just one gulp! Today, we don't call them monsters; instead, we refer to them as giant marine reptiles. As far as we know, none of them are alive today. They are all now probably extinct.

# Amazing Creations

The giant marine reptiles gave glory to God.

Where did the giant marine reptiles come from? The answer is simple. God made them. He made them on the fifth day of creation: on the same day He made all the other wonderful swimming creatures. If you read the Bible carefully, you will find many places where the Scriptures mention the great creatures of the deep. Isaiah 27:1 talks about "the dragon who lives in the sea." Psalm 74:13 reads, "You divided the sea by Your strength; You broke the heads of the sea monsters in the waters." In Psalm 104:26, a creature named **leviathan** (luh vye' uh thuhn) is described as playing in the deep waters. I guess a huge reptile would need deep water in which to play, wouldn't it? In the book of Job, chapter 41, God Himself describes leviathan as terrible and frightening. Why would God make big, scary sea monsters? That answer is also in the Bible. Psalm 148:7 reads, "Praise the Lord from the earth, Sea monsters and all deeps." The monsters of the deep give their praise to the Lord. They show us His might, His strength, and His power!

*Tell someone about how the first giant marine reptile was discovered and how the Bible mentions such creatures.*

# Four Saurs

We are going to study some of the giant marine reptiles whose fossils scientists have discovered. Keep in mind, however, that what we are about to discuss is not certain. It is just what scientists currently believe. You see, scientists have never actually seen these creatures. They have only seen these creatures' bones. There is not much you can tell about something from its bones, yet scientists have made guesses about what these creatures looked like and how they behaved based on these bones. Of course, scientists are not always right when they make such guesses. You need to keep that in mind.

The fossil record tells us that there were many giant marine reptiles, but I want to concentrate on four basic types: **nothosaurs** (noh' thuh sawrz), **mosasaurs** (moh' suh sawrz), **ichthyosaurs**, and **plesiosaurs**. Before we talk about the differences between these animals, let's talk about their similarities. First, they were probably reptiles. This means that they were air-breathers, just like sea turtles and sea snakes. Scientists believe that they were ectothermic, so they probably would have preferred warmer waters. Their offspring hatched from eggs. Their skin had scales. They were meat-eaters and were probably ferocious hunters. Their mouths were full of sharp, pointed teeth that were designed to catch and hold slippery fish. They seemed to be great swimmers. With the possible exception of the nothosaurs, they spent most (if not all) of their time in the water. Since their fossils are sometimes found in groups, they may have lived in herds.

# The Nothosaurs

Nothosaur means "spurious lizard," and while it isn't a true lizard, the nothosaur did look like a lizard with a long neck and tail. Imagine a long-necked, long-tailed lizard with four webbed feet. Imagine a smile on its face, revealing sharp, pointy teeth. Now in your imagination make the lizard bigger and bigger until it is 10 feet long. If you can imagine that, you have a pretty good idea of what a nothosaur probably looked like.

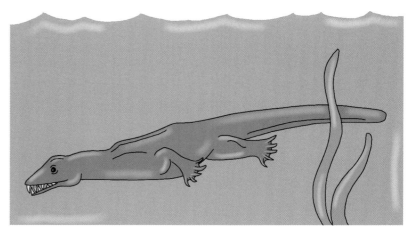

Its webbed feet helped a nothosaur feel at home in the water.

The nothosaurs were unusual because they had feet with long, webbed toes. These specially designed feet enabled them to be able to walk around on the beach and also swim and dive gracefully. Think about a duck's feet. The fact that they are webbed allows the duck to walk on land as well as swim effectively. It was probably the same for the nothosaur. It probably also swished its long tail back and forth to push itself through the water.  Since nothosaurs could both walk on land and swim, no one really knows whether they spent most of their time on the beach or in the water.  Perhaps the nothosaurs lived primarily in the water but enjoyed sunning themselves on coastal rocks, much like the lizards of today enjoy basking in the sunshine.

Nothosaur babies hatched out of eggs buried in sand in the same way that turtle eggs hatch.  Baby nothosaurs probably fled for the safety of the water as quickly as possible to keep from being eaten by hungry predators, especially hungry birds or pterosaurs.  We don't know whether or not nothosaurs cared for their offspring.

Because a nothosaur could move around in and out of the water, it could catch small land animals along the shore or it could dive into the water to catch fish.  Since its nostrils were located on the top of its long snout, it did not have any trouble breathing as it went cruising along the top of the water.

Nothosaur fossils have been found in many parts of the world.  While some kinds of nothosaurs have been found to grow as long as 20 feet, others like the **lariosaurus** (lar' ee oh sawr' us), a kind of nothosaur discovered in Europe, was only about 2 feet long!

*Describe what you remember about nothosaurs.*

# The Mosasaurs

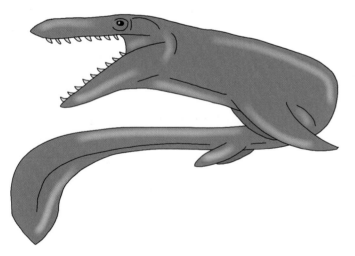

Mosasaurs had snake-like bodies, tiny flippers, and paddle-shaped tails.

The mosasaurs were less lizard-like than the nothosaurs.  In fact, they had slim bodies and resembled sea snakes more than lizards.  Instead of legs, mosasaurs had flippers, like sea turtles.  Their flippers were small and weak compared to their size, so they had to rely on their long, skinny tails to push them through the water.  A mosasaur's tail was wide at the end and was shaped something like a paddle.  This gave it a bigger surface with which to push against the water as it swished from side to side.

Small mosasaur fossils have been found, indicating that some grew "only" 10 feet long.  However, it appears that most of them were much bigger, some growing up to 50 feet long!  Those big bodies had big, powerful jaws, and the inside of the mouth was crowded with cone-shaped teeth.

Can you imagine a mosasaur sitting in a dentist chair with its mouth open? Do you think the dentist would say, "Open wider, please?" Well, if he did, you might be surprised to learn that the mosasaur could do as the dentist requested! You see, its lower jaw was hinged, similar to a snake's jaw, so it could unhinge its lower jaw and swallow prey even bigger than its mouth. Because the mosasaur had this special feature, scientists think

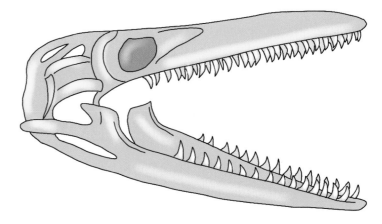

Mosasaurs had large teeth and hinged jaws so their mouths could open wide.

that it swallowed its food in one big gulp, never chewing. Why would an animal that doesn't chew need such a big, powerful mouth? One reason was because mosasaurs enjoyed eating animals with tough shells. A mosasaur probably used its jaws in much the same way as we use nutcrackers to break the tough outer shells of nuts to get to the soft meat inside. Fossils of clams, ammonites, and sea turtles have been found with the clear marks of mosasaur teeth on them. But a mosasaur didn't survive on shelled animals alone. Indeed, mosasaurs ate fish, other mosasaurs, and even *birds*. The mosasaur must have crept up under the birds while they dived for fish and ambushed them, swallowing a whole bird with one big bite.

Even though the mosasaur was a predator, it appears that at times it was prey as well. The teeth of sharks have been found embedded in the fossils of mosasaurs, so sharks must have seen mosasaurs as a tasty menu item. Of course, it is hard to say for sure, because the sharks may have been scavenging on the bodies of mosasaurs that had died for other reasons. A mosasaur would have been a hard kill for a shark, not only because it would have fought back with its huge mouth, but also because it had tough scales that acted something like armor plating.

Since mosasaurs didn't have legs, they probably did not ever leave the water. Their flippers would not support the weight of their bodies. This presents a question about how they reproduced. Reptiles lay eggs, right? If mosasaurs couldn't lay eggs on land like the nothosaurs, where could they lay them? If their eggs were laid directly in the ocean waters, the developing babies inside them would drown. So most researchers have come to believe that the mosasaur kept its eggs inside its body until the eggs hatched and then gave birth to live young. Do you remember that sea snakes do this very same thing? Do you remember what word we use to describe this? That's right: ovoviviparous.

In South Dakota, a mosasaur fossil was found that backs up this belief. Inside the mosasaur's abdomen were four mosasaur babies. However, if you know where South Dakota is, then your question should be "What was a mosasaur fossil doing in South Dakota, where there is no ocean?" Good question. Would you believe that mosasaur fossils have been found all over the entire world, on every single continent, even Antarctica? Most of the states in the United States have mosasaur fossils, even in the desert! Interesting, huh? What are some reasons you can think of for this? Keep thinking; we'll explore a possible answer in a little bit. First let's take a look at the next group of sea serpents.

*Tell someone what you have learned about nothosaurs and mosasaurs.*

# The Plesiosaurs

Have you ever heard this joke: "What do you get when you cross a cat with a parrot? A carrot!" Or how about this one: "What do you get when you cross a sheep and a kangaroo? A woolly jumper!" Those are pretty funny. What about this one? "What do you get when you cross a snake with a sea turtle? An **elasmosaurus** (eh laz'moh sawr' us)." Well, that one probably didn't sound funny, but it's actually kind of true. An elasmosaurus looks like a snake that was forced into the body of a turtle and then lost its shell. The elasmosaurus was a kind of plesiosaur. Plesiosaurs were marine reptiles with really super-long necks. The plesiosaurs were part of a group of marine reptiles called **plesiosauria**, which also includes the short-necked **pliosaurs** (plee' uh sawrz). We'll learn more about pliosaurs later.

Of course, you know that plesiosaurs, and elasmosaurus in particular, were not really snakes and turtles mixed together! No, they were special creatures designed by God with wonderful features that allowed them to move around like no other sea creature. Plesiosaurs had four flippers that they used like wings in the water. You might say they could "fly" through the water! Do you remember another reptile from the last lesson that could fly through the water? Of course: the sea turtle! However, sea turtles flap only their front flippers. Researchers believe that each of a plesiosaur's four flippers could be controlled separately, and if this were true, a plesiosaur would be able to turn completely around in one place like a top spinning in place on the table.

A plesiosaur could have changed direction at any time with grace, performing incredible underwater acrobatics. This is especially amazing since some plesiosaurs were enormous. Elasmosaurus was almost 50 feet long. Maybe this was the creature called leviathan in Psalm 104:25-26 that was said to "play" in the deep! *"There is the sea, great and broad, in which are swarms without number, animals both small and great. There the ships move along, and leviathan, which You have formed to sport in it."* Of course, leviathan might have looked more like the pliosaur that we'll learn about later. At the end of this lesson, you can do an exercise to figure out what you think the leviathan looked like. So why could this huge animal swim so gracefully? It's because of how God designed its hand bones.

## Try This!

Feel your hand and try to guess how many bones you have in each finger. Did you say three bones in each finger? If so, you were correct! Your thumb has only two bones, but your other fingers each have three. Since you have three bones in each finger, your fingers are flexible; they can bend.

At each bending place, you have a joint.  How many finger joints do you have?  Count them now.  You have fourteen in each hand.

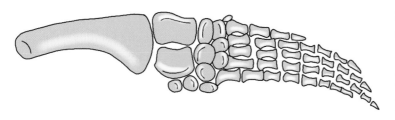

This drawing of fossilized flipper bones shows you that plesiosaurs had lots of bones in their flippers.

Plesiosaurs had many, many more bones in their "fingers."  They had up to twenty-four different bones in each finger.  Try to imagine what it would be like to have twenty-four bones in each of your fingers.  They would be very long and very flexible wouldn't they?  How many joints would a plesiosaur have?  Can you figure it out?

Plesiosaurs had an unusual number of bones in their fingers, but that was not the only place they had extra bones.  In case you haven't realized, plesiosaurs had very long necks!  In an elasmosaurus, more than half of its length was its long neck.  The average reptile, like a lizard or turtle, has only five to ten neck bones, but elasmosaurus had more than seventy.

That long neck does cause us to ask some questions.  How did a plesiosaur balance its long neck when its body and tail were so short in comparison?  To see what I mean, try to hold a broom horizontally out in front of you.  What do you feel?  Does it feel like you are going to fall forward?  Do you think it would be even harder to hold the broom in front of you while you were swimming?  Another question is, "How did a plesiosaur breathe through its long neck?"  You see, the neck was very narrow compared to the rest of the body.  Not much air would be able to move through it at one time, and it would take a long time for the air to pass all the way from the nostrils to the lungs and back out.  Try breathing through a coffee stirrer or a thin straw for a few minutes and you will understand the problem.  At this time, only God knows the answers to these and many other questions about the plesiosaur.  Perhaps God will allow scientists to find the answers one day.

Well, we've talked about the long neck or the "snake" part of the plesiosaur, but what about the "turtle" part?  First of all, the body of the plesiosaur may have been shaped similarly to that of a turtle, but it did not have a shell.  It did, however, share one of the unusual features of a turtle.  It had something called **gastralia** (gas' truh lee' uh), or belly ribs.  These bones are pointed out on the elasmosaurus skeleton drawing below.

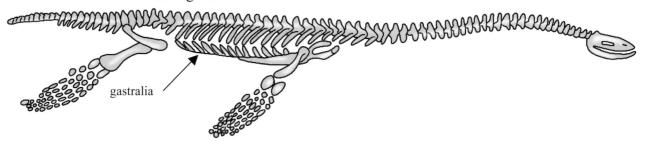

gastralia

Notice that the neck of this elasmosaurus is longer than the rest of its body!

Turtles also have these ribs, embedded in their plastrons. Crocodiles have them, too. Turtles spend a lot of time resting on their bellies, and crocodiles do, too. Perhaps the fact that plesiosaurs have gastralia tells us that they also spent time resting on their bellies.

This idea makes sense because plesiosaurs have been found with fossils of the kind of food that would have been available from the ocean floor: shellfish like oysters, mussels, clams, and snails. But plesiosaurs couldn't crack shells like the mosasaurs did. The teeth of the plesiosaurs were not designed for cracking shells. Instead, they probably swallowed them whole. In the belly of a plesiosaur were "stomach stones," which are called **gastroliths** (gas' truh liths). These stones moved around in the plesiosaur's stomach and cracked or crushed the shells of the animals it ate. Did you know that some birds swallow stones to help them digest food as well?

These stones are gastroliths that were found with marine reptile fossils. The golf ball in the center is there to give you an idea of how big the gastroliths are.

You can tell gastroliths from other stones because they are smooth and polished from rubbing against each other in the belly of the plesiosaur. Plesiosaurs probably swallowed rocks made of granite and quartz, which means the rocks were very hard. Why do you think that is? I suppose that God gave plesiosaurs the instinct to search for just the right stones. Quartz is often shiny and easy to see sparkling in the water. Perhaps that's why plesiosaurs chose it. Plesiosaurs had large amounts of gastroliths in their stomachs. One plesiosaur fossil found in South Dakota had 253 gastroliths weighing a total of 29 pounds! Aren't you glad you don't have to swallow rocks before you eat?

Plesiosaurs may have gone hunting for their food as well as feeding from the sea bottom. Their heads were very small and, as you know, very far away from their bodies. Some scientists have suggested that their small heads may have been camouflaged to look like fish, enabling their long necks to maneuver the head right into the middle of a school of fish with out being detected. Their eyes were located at the top of the head so they could look upward toward prey swimming overhead. Their sharp, curved teeth could have been called "ratchet teeth" because they curved backwards, allowing the prey to move in only one direction, toward the stomach. Plesiosaurs had unusual nostrils. It seems that water was able to pass right through them and exit out the top of the head. No one really knows what purpose this might have served. Some say this allowed them to smell or maybe even taste the water to aid them in hunting.

*Explain in your own words all you can remember about the plesiosaurs.*

# The Pliosaurs

The long-necked plesiosaurs were not the only members of the group of reptiles called plesiosauria. In fact, there were reptiles similar to plesiosaurs, but they did not have really long necks. They were the pliosaurs. The pliosaurs had short necks and huge heads!

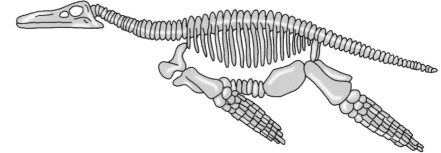

This drawing of a pliosaur skeleton shows you that pliosaurs were similar to plesiosaurs, but they had much shorter necks.

They also had four long flippers that they used for swimming. Their slimmer, streamlined bodies and shorter necks caused little water drag, enabling them to swim faster than the odd-shaped, long-necked plesiosaurs. This additional speed probably made them better hunters. Their teeth were more cone-shaped, like those of the mosasaurs, great for crunching hard-shelled animals.

If size has anything to do with it, the hungriest pliosaur might have been the **kronosaurus** (kroh' nuh sawr' us). Kronosaurus was 30 feet long with a head that was 9 feet long. To truly appreciate the size of that head, you probably need to see it.

## Try This!

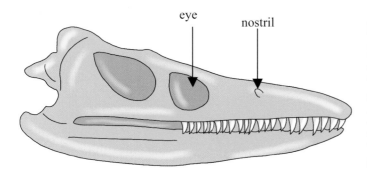

Get a tape measure and some chalk and head outside to find a large place to draw. Sketch a triangular-shaped head that's 9 feet long and about 3 feet wide. On the two long sides of this triangle draw rows of teeth. Make each tooth about the size of a banana. Place nostrils on top of the snout and eyes about two-thirds of the way back on the head. Use the drawing on the left to help you.

All members of the group called plesiosauria, both plesiosaurs and pliosaurs, including kronosaurus, were probably ovoviviparous, like the mosasaurs. Some think that the adults cared for their young, but there is no solid evidence for this idea. Remember, we have to hypothesize everything we know about the animal's behavior from fossilized bones alone. This makes it very hard for paleontologists (pay' lee on tahl' uh jists), scientists who study fossils, to know exactly what the animals looked like and how they lived.

*What do you remember about the pliosaurs?*

# The Ichthyosaurs

Now we will learn a little about Mary Anning's "sea monster" — ichthyosaurus. If you saw an ichthyosaur swimming in the ocean, you would probably think that it was a large fish, or maybe even a dolphin; that is, until you caught one and looked carefully at the bones. Then you would see that what appeared to be fish fins were really limbs — arm, hand, and finger bones in the front, and leg, foot, and toe bones in the back — all hidden inside of fin-like flippers. You might say that an ichthyosaur was a lizard skeleton hidden inside a fish body. That's where it got its name. Ichthyosaur comes from two Greek words that mean "fish lizard."

Ichthyosaurs probably looked a lot like dolphins.

When researchers first studied an ichthyosaur's fossilized bones, they were not sure exactly what it looked like when it was alive. The most puzzling part was its tail. The tail bones did not seem to be arranged like the bones of similar marine reptiles. Fortunately, it wasn't long before a new group of ichthyosaur fossils were discovered in Germany. These fossils were preserved in such fine sediment that the fossils were almost perfect, with few bones disturbed or missing. When a fossil is found with the bones present in their proper places, almost as if they were all still attached to each other, we say that the fossil is **articulated** (ar tik' yoo lay' ted). The articulated bones of these ichthyosaur fossils held another clue as to what ichthyosaurus really looked like.

Notice that the imprint on this fossil shows what the tail looked like.

You see, the fossil had a dark imprint which was an outline of the entire body. Even though the soft parts had rotted away long ago, they had left their imprint behind. For the first time, scientists could see clearly that the tail of ichthyosaurus was shaped like the tail of a fish! And on its back, also like a fish, was a dorsal fin. However, even though the ichthyosaurus tail looked like a fish tail, again, the bones on the inside showed that the ichthyosaurus was not a fish. The backbones of reptiles and fish are quite different, and ichthyosaurus clearly had a reptile's backbone.

Because ichthyosaurs were shaped like fish, we think they probably lived like fish or dolphins. Their streamlined bodies were well-suited to cruising through the water very quickly and diving deep in the ocean to hunt for food. We think they went deep into the sea because squid have been found fossilized in their stomachs. Squid tend to live deep in the ocean. Also, they usually had large eyes. Perhaps God provided them with large eyes so that they could see better at deeper depths. In fact, the largest ichthyosaurs had the biggest eyes in creation — 12 inches across! That's as big as a dinner plate!

Probably the most famous ichthyosaur fossil ever discovered is a female giving birth to a baby ichthyosaur. The mother and child died at the very moment of birth, the baby still in the process of being born. But what kept hungry scavengers from eating their carcasses? How did their bones remain articulated (Do you remember what that means?) rather than being scattered by the currents of the water? The answer is that the ichthyosaurs were probably covered quickly with sediment during an underwater mudslide or a volcanic eruption. In fact, it was probably this sudden catastrophe that actually killed them.

## Try This!

It's rare for animals like ichthyosaurs to become fossilized. Do you know why? Let's find out. You will need a seashell, a Cheerio (or similar piece of breakfast cereal), two glasses, and a container to hold some water and dirt. The seashell represents a sea creature like a clam. The Cheerio represents an animal like an ichthyosaur, because it has pockets of air in it, just like an ichthyosaur does. Place the Cheerio in the bottom of one glass and the shell in the bottom of the other. Cover each with a small amount of water. In the container, mix a little water with the dirt until you have made some nice, sloppy mud that will pour. Carefully pour the mud into each of the two glasses. What happens to the shell? What happens to the Cheerio? Watch the Cheerio for a few minutes. What is happening to it now? Why do you think that animals like ichthyosaurs don't often fossilize? If you said that they float, you are right. Scavengers are attracted to floating bodies, so they often get eaten before they have a chance to fossilize. Usually, an animal needs to be trapped beneath deep layers of mud (sediment) to become a fossil.

Many aquatic animals float when they die, making it very hard for them to fossilize.

Still, however, some ichthyosaur fossils have been found. One of the largest finds was in the state of Nevada. There, thirty-seven **shonisaurs** (shon' ih sawrz' – a type of ichthyosaur) were found lying near one another. Some scientists think they were beached, like whales. However, that doesn't make a lot of sense. Beached animals rarely form fossils. Instead, they usually are eaten by scavengers and decay away. In addition, the shonisaurus fossils were mostly articulated. This indicates the dead animals did not spend a lot of time exposed to currents, tides, or other forces that would tend to scatter their bones.

How, then, did all of these shonisaurus fossils form? The only reasonable explanation is that the animals were buried in sediment (mud) soon after they died. That would keep scavengers from eating them, and it would keep currents and tides from breaking the skeletons apart. Can you imagine how much mud it would take to cover that many large animals? And remember, they would have had to be covered quickly and completely so that they would not be eaten by scavengers or decay. What kind of a catastrophe could have moved that much mud quickly?

# The Deluge

The Bible records an event that could have covered all of those shonisaurs quickly. In Genesis 6-8, a worldwide flood is described. Some call this the Great Deluge, which just means the great Flood. At this time, the "fountains of the deep" broke open, as the Bible says. Water from under the earth and under the sea broke forth and covered the entire earth. This event would have easily caused enormous mudslides. If you've never seen a mudslide, check out this picture of a mudslide that happened not long ago when it rained too much in Wyoming. If too much rain can cause this, imagine what happened when the "fountains of the deep" broke open! The world was subjected to mudslides of enormous proportions, and many creatures, great and small, were covered in mud.

Notice that the mud from this mudslide trapped three cars that were traveling down the road at the time of the mudslide.

The Great Deluge can also explain why we find the fossils of giant marine reptiles all over the earth, even where there is no water today. After all, when the waters covered the entire earth during the great Flood, the giant marine reptiles could have gone anywhere. If they happened to meet with a giant underwater mudslide as the floodwaters spread, the mud would have quickly encased them and deposited them wherever the mudslide stopped — perhaps even in your own backyard!

# What Do You Remember?

Who found the first fossil of a giant "sea monster?"  What four kinds of large sea reptiles did we discuss?  Which of those four might not have spent all of its time in the sea?  Which two animals is a plesiosaur like, and how is it like them?  What did the plesiosaur eat to aid in chewing its food?  How do we know what an ichthyosaur looks like?  How is an ichthyosaur different from a fish?  Which of the four types of giant marine reptiles was shaped a lot like a snake and had tiny flippers?  What is one explanation for why we have so many sea creature fossils all over the earth?

# Notebook Activities

After recording the fascinating facts you learned about in this lesson, draw an underwater scene with a nothosaur, a mosasaur, a plesiosaur, and an ichthyosaur.  Label each animal, and then, write a few facts about each of these giant marine reptiles.  Pages are provided in the *Zoology 2 Notebooking Journal.*

**Older students**: Many people say that leviathan described in the Old Testament is a whale or a large crocodile.  Read about leviathan in Job 41, paying close attention to the description.  Then, I want you to use Venn diagrams to compare and contrast leviathan with these animals.  A Venn diagram is a wonderful tool that helps you compare and contrast different things.  The drawing on the right is an example of what a Venn diagram looks like.  In your notebook make your own Venn diagram that you will use to

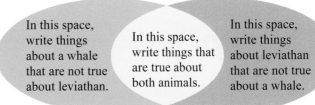

compare and contrast a whale with leviathan.  If you have the *Zoology 2 Notebooking Journal*, a page is provided there for you to use for this activity.  Write details about leviathan in one oval of the Venn diagram and details you know about whales in the other.  If they have any similarities, put them in the center where the ovals overlap.  Do this again to compare and contrast a crocodile with leviathan.  Finally, draw a picture of what you think leviathan looked like.

# Experiment

Animal fossils are made when animal bones are preserved in sediment (mud or dirt).  The sediment preserves many of the details of the bones.  The better preserved the bones are, the easier it is for paleontologists to study and learn about the creature.  But which sediments preserve details nicely?  Would the sediment around your home preserve the details of a fossilized sea creature?  Let's do an experiment to find out.

There are three basic types of sediment: sand, silt, and clay. The main difference among these types of sediment is the size of the particles that make them up. Sand particles are much larger than the particles that make silt and clay. So we will do an experiment to see whether large particles (sand) or smaller particles (silt and clay) preserve details better. At the same time, we will test your soil to see if it would make good fossils.

**You will need:**

♦ A "Scientific Speculation Sheet"
♦ Three small plastic disposable containers
♦ A cup of sand
♦ A cup of clay (Real clay is preferable, but modeling clay is okay.)
♦ A cup of soil from your yard (Collect your soil at least 12 inches below the surface.)
♦ A detailed shell of some sort (It should have ridges or other specific features. If you can't get a shell, you can use a toy that has detail in its shape.)
♦ A craft stick or plastic spoon for stirring
♦ Two cups plaster of paris
♦ A cup of water
♦ A disposable container for plaster
♦ A permanent marker

1.  Use a "Scientific Speculation Sheet" to record your hypothesis of whether sand or clay will make the better fossil.
2.  Put the sand in one container, the clay in another container, and your soil in the last container.
3.  Moisten the contents of all three containers.
4.  Prepare the plaster by mixing it with the water according to the instructions on the package. Stir it well to break up any lumps. The plaster should be the consistency of glue and should pour easily.
5.  When the plaster is well mixed, carefully press the shell into the sand, remove it, and quickly pour one-third of the plaster over the impression left in the sand, covering it completely.
6.  Wash the shell and repeat the process in the clay.
7.  Wash the shell and repeat the process with your soil.
8.  Allow the plaster to completely dry. Once the plaster is dry, you will have a **cast** of the shell's imprint in the sand. This is one way fossils form. The creature makes an imprint in sediment, and then when different sediments fill the impression and harden, they make a cast of the impression.
9.  When the plaster is dry, carefully remove the plaster from the sand container. You may have to break the container to get it out, but be careful to not break the plaster cast.
10. Wash away any sand that is clinging to the plaster cast and use the marker to label the back of the cast, SAND. (NOTE: Don't put the sand or clay down the drain!)
11. Remove the clay cast, clean it, and label it, CLAY. Examine each cast. Does the cast made in sand or the cast made in clay look more like the real thing? That will tell you whether sand or clay is better for forming fossils. Was your hypothesis correct?
12. Now remove the cast from your soil and label it, SOIL. How does your soil measure up?

# Lesson 6
# Fish

Now that you know about aquatic mammals, reptiles, and amphibians, it's time to launch into a study of God's fabulous fish! When God created fish, He was especially imaginative. There are beautiful fish, ugly fish, huge fish, teeny-tiny fish, fish that have strange contraptions hanging off the front of their heads, and fish that are so brightly colored we catch them and put them in aquariums just for people to enjoy. There are so many, many different kinds of fishes. Some look like rocks, while others look like snakes. Some have huge mouths

There is an amazing variety of fish in God's creation.

and sharp teeth, and others can actually fly! The variety of fishes in the sea is so great that I couldn't possibly tell you about each one. Instead, I will spend time talking about the features most fish have, and along the way, I'll tell you about some of the different kinds of fish in creation.

Many people, grown ups and children alike, aren't sure whether to say "fishes" or "fish" when they are talking about more than one fish. Actually, both are correct. A group of fish from the same species should just be called "fish." However, when you are talking about two or more different *species*, you should say "fishes." So if you see two guppies, you say that you are seeing two fish. If you see a guppie and a catfish, you say you are seeing two fishes. Interesting, huh?

Although it doesn't look like one, this sea horse is, indeed, a fish!

There are three main kinds of fishes: **bony fishes**, **cartilaginous** (kar' tih laj' uh nus) **fishes**, and **jawless fishes**. Bony fishes make up almost all the fishes in the sea. Cartilaginous fishes are sharks and rays; we'll study them in the next lesson. The last kind, jawless fishes, is a strange group indeed. We'll study them in the next lesson as well.

So what makes a fish a fish? Can you take a guess? Is it scales? Well, no. Sea snakes have scales, but they aren't fish. Is it that they lay eggs? No; sea turtles lay eggs but are not fish. Is it their shape? No, again. A seahorse is a fish but isn't shaped like a typical fish. So, how do you know if you are looking at a fish? Scientists say that if it has fins for swimming and gills for breathing, it's a fish! Sharks are fish, but dolphins are not, even though dolphins look a bit like sharks. Sharks breathe with their gills, but dolphins breathe with their lungs.  A sea snake looks almost identical to an eel, but an eel usually has fins and breathes with gills; so an eel is a fish, while a sea snake is a reptile.

Fish are ectothermic (cold-blooded), moving more quickly in warm water than in cold water.  However, they actually prefer cooler water.  Do you want to know why?  It is because cooler water contains more oxygen, making it easier for them to breathe.  In fact, some fishes can't survive in warm water because of the lower level of oxygen.  Most fishes can withstand extremely low temperatures, even freezing cold water.  In fact, they congregate in enormous numbers in the freezing waters of the Arctic.  Even though they are ectothermic, some predatory fishes, such as tuna and sharks, have muscles that actually produce heat.  This warms their bodies a bit, making them able to move more quickly, even in cooler waters.  This aids them in catching prey, because the faster swimmer usually gets the food.

# Bony Fishes

Have you ever seen the symbol of a fish on the back of someone's car?  That symbol is called an "*ichthus*," which is a Greek word for "fish."  Most fish are in the class **Osteichthyes** (ahs tee ik' theez), which means bony fish.  Now look at the word "osteichthyes."  If "osteichthyes" means "bony fish," and you know that the Greek word for fish is "*ichthus*," what do you think the Greek word for "bone" is?  That's right!  The word "*osteon*" is a Greek word that refers to bone.  As the name indicates, osteichthyes are fishes that have bones.  Although sharks and rays are fishes too, they don't have bones; neither do the jawless fishes.  Now don't misunderstand.  Sharks, rays, and the jawless fishes have *skeletons*, but those skeletons are not made out of *bone*.  You will learn what I mean in the next lesson.

Bony fishes have scales that cover their bodies.  Each scale is attached to the skin, but not very tightly, for it can come off easily in your hand.  On top of the scales is a thin coat of slime, which the skin produces to protect the fish from parasites.  It also helps the fish swim more easily through the water.  If you have aquarium or pond fish, you shouldn't pick them up with your hands because you can wipe off this protective film very easily.  Without this slime, the fish is vulnerable to being attacked by all the parasites floating around in the water.  This

This close up makes the fish scales very easy to see.

is why many fish die soon after they are thrown back in the water after being caught by a fisherman.  The fisherman couldn't help but rub off slime when he unhooked the fish to toss it back in the water, and that left the fish unprotected.

Fish scales are kind of like fingernails; they grow bigger as the fish grows larger. As the scales grow, they form rings called **circuli** (sir' kyu lye). If you were to take a scale from a fish and count the circuli, you would have a good idea of how old the fish is. This is similar to how certain turtles add growth rings to their scutes.

Have you ever noticed the amazing variety of mouths God gave fish? God gave each type of bony fish a mouth that is best suited for getting the food it eats. Some fish have mouths that angle upward so that they can easily feed on food that floats at the surface of the water. Bottom-feeding fishes have mouths angled downward for feeding on the clams and other creatures that lie on the bottom of the ocean. Some fish are fearsome predators whose mouths are filled with sharp teeth. Fish that inhabit coral reefs and other tropical areas often have thin, long snouts with small mouths that can reach into small cracks, crevices, and slits in the reef walls.

Notice the incredible variety of mouths on these fishes.

# Grand Gills

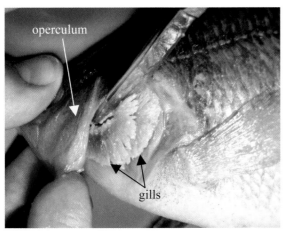

In this picture, the operculum has been pulled back to reveal the gills.

Fish gills are located near the head, on either side of the fish's body. If you happen to have fishes in an aquarium, look closely to see if you can find the gills. If they are hard to see, it's because over the gills of most bony fishes is a flap called the **operculum** (oh per' kyoo luhm). The operculum opens and closes every time the fish breathes. Gills of sharks and rays don't have this protective operculum over their gills. Instead, they have gill slits, which I will cover in the next lesson.

If you have fish in an aquarium or bowl, watch them breathe. Can you tell whether the water is going in the gills or out the gills? Do you notice anything about the fish's mouth? If you have a chance to watch a fish closely, you may notice that when it breathes, its mouth and operculum open and close at the same time. It works like this: Water goes into the mouth and passes over the blood vessels in the gills, where oxygen from the water is absorbed into the blood vessels. As oxygen goes into the blood vessels, the blood also releases a waste gas called carbon dioxide.  The carbon dioxide goes into the water that is passing over the gills, and then the water exits through the operculum.

Just like a fish, you need to breathe in oxygen and breathe out carbon dioxide.  When you breathe in, air goes into your lungs, and the oxygen in the air enters blood vessels in your lungs.  At the same time that oxygen is going into your blood vessels, carbon dioxide is released by the blood vessels.  When you breathe out, that carbon dioxide leaves your body.  So, you breathe in air that has oxygen in it, and you breathe out air that has carbon dioxide in it.  Fishes take oxygen-filled water into the mouth, and water that has carbon dioxide in it leaves from the operculum.

# Fabulous Fins

In order to swim well, a fish must have good swimming equipment.  God gave all the animals we have studied so far great swimming equipment, like flippers and tails that move them through the water.  Fishes also have great swimming equipment called **fins**.  The fins enable a fish to move, and sometimes, if the body is shaped right, to move very, very fast.  Do you remember that animals that have the ability to get from place to place by swimming are called nekton?  They are different from plankton, which mostly drift with the currents.

A fish swims in a way that is similar to how a snake wriggles through the grass.  Back and forth its body twists, as it moves through the water in the shape of an "S."  While its body is twisting in this side-to-side fashion, its tail pushes it forward, and its fins steer it around.

Look at the fish diagram on the right.  Can you find the **pectoral fin** and the **pelvic fin**?  These two fins are the only ones that are paired, meaning there is one like it on each side of the fish's body.  Pectoral and pelvic fins are attached to muscles inside the body.  When the fish's

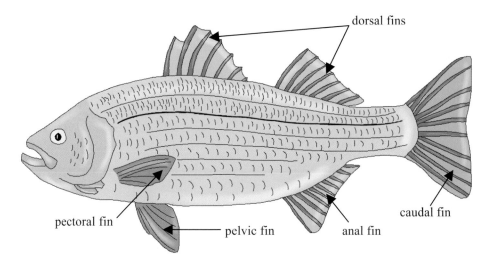

dorsal fins

pectoral fin

pelvic fin

anal fin

caudal fin

muscles squeeze and relax, the fins move away from the body and back in toward the body.  Having these four fins helps the fish move in the direction it chooses.  Without the pectoral and pelvic fins, the fish would not be able to go where it wants to go.

Now look for the **dorsal fins**.  There are two of them on this fish.  These fins, along with the **anal** (ay' nuhl) **fin**, keep the fish from rolling side to side, falling over this way and that, as it swims.

In other words, the dorsal and anal fins give the fish balance.  The dorsal fin is also somewhat of a weapon for the fish.  Most have sharp spines; some are even filled with venom.

An untrained fisherman will learn about this dorsal fin weapon rather quickly.  When pulling a fish out of the water, a fisherman must carefully fold the dorsal fins back down over the back of the fish when he grabs the fish to unhook it.  If he doesn't, he will be cut through the skin by the sudden thrust of the dorsal fin's spines into his hand.  Once the dorsal fins are safely out of the way, the fish is easy to handle.

Now find the **caudal** (kaw' duhl) **fin**.  The caudal fin is what we call its tail.  It is used to thrust the fish forward in the water while it swishes from side to side.  The shape of the caudal fin often determines how fast the fish can swim.  Caudal fins that are forked and thin like knives (like those of swordfish and sharks) belong to the fastest swimmers, while caudal fins that are rounded and thick belong to the slower swimmers.  While the shape of the caudal fin is important, you'll learn in a moment that the shape of the body is the most important thing when it comes to speed.

# Shaping Up

Which fish swims faster, a tuna or an angelfish?  The key to fast swimming is the shape of the body.  The smoother and more rocket-shaped, the faster a fish can glide through the water.  This is because streamlined body shapes reduce drag.  Remember, water drag slows things down as they try to move through the water.  With the right shape, drag can be reduced by quite a bit.  Let's look at some of the shapes God gave His fishes and how each shape serves the fish well.

The fusiform shape of this shark makes it a very fast swimmer.

Notice the shape of the shark in the picture above.  It is called a **fusiform** (fyoo' zuh form) shape, and it is defined as a rounded, bullet shape where both ends of the fish taper to a point.  This is the best shape for speed.  Fishes with fusiform shapes can chase and catch prey quickly.  They can also escape predators.  Many live in the open ocean and swim constantly.  Examples of fusiform-shaped fishes are sharks, marlins, swordfish, and tuna.

Have you ever seen an eel?  If you haven't, look at the picture on the next page.  The long, ribbon-shaped body of an eel illustrates another fish shape.  Not surprisingly, this kind of shape is called **eel-like**.  Eels and fishes like them are slow swimmers.  They are usually secretive, hiding in cracks, caves, and crevices. Their bodies are designed so that they can spring out quickly to capture prey, and then return just as quickly to their hiding spot. They tend to remain still until prey swims by, then quick as lightning, they lunge out and seize their victims. Examples of eel-shaped fishes are gunnels and eels.

The shape of an eel is perfect for hiding in crevices.

The flounder pictured below has a **depressed** shape. Can you think why a fish would have such a shape? Because it likes to lie on the bottom of the ocean or lake in which it lives. Fishes that have a depressed shape are not sad at all, but happy to be shaped like this, because it allows them to hide from predators. A lot of these fishes are actually able to change colors to match the different colors of the ocean floor. All depressed fishes can see out of both eyes when they are lying down. A depressed fish has both eyes either on top or on one side of its body. Examples of depressed fishes are batfish (both eyes on top of its body), rays (both eyes on top), and flounders (both eyes on one side).

A flounder has both of its eyes on one side of its depressed body.

If a fish does not have a fusiform, eel-like, or depressed shape, it might have a **compressed** shape. These fishes have flat bodies, with one eye on each side. Think about how this is different from a depressed shape. A fish with a depressed shape is flat, but it has both eyes on one side or on top of its body. Compressed fishes have an eye on each side of the body. They swim upright and can be very thin. When seen from the front or back, they can almost disappear like a thin line. You can find these fishes in reefs where their flat bodies can easily dart in and out amongst the corals, sponges, and rocks, keeping hidden from predators. Examples are angelfish, surgeonfish, and butterfly fish.

Now these are not all the shapes that fish can have, but they are the more common ones. There are some fish that have a combination of these shapes, and some, like the sea horse, that have their own unique shape. It must have been a lot of fun to design all the shapes, features, and colors of the fish in the ocean!

*Describe the different fish shapes I just told you about.*

Notice how thin this compressed fish looks from the front.

# Defense

If a fish doesn't have the ability to swim fast, it often has some sort of defense. Do you remember what defense is? It's a design God gave animals to protect themselves. Fishes have several defenses. Let's examine a few.

Some fishes use **camouflage** (kam' uh flahj) as a defense. When a fish is camouflaged, it looks like the environment in which it dwells, so it is difficult for a predator to see it. A flounder, for example, is flat (depressed) and is often colored to match the sand or dirt on the ocean floor. Rays also have a depressed shape, and they lie on the bottom of the ocean, usually with colors that are mottled, spotted, and freckled, so that they look like the ocean floor. Many fishes have coloration called **counter-shading**. This means that the body is light on the bottom and darker on the top. Because of this, when a predator is looking up at the fish swimming above, the

This flounder blends in with the sand in which it lies.

predator may not notice the fish, as its light belly allows it to blend in with the light streaming down from the surface of the water. If the predator is looking down at the fish swimming below, the fish's darker top allows it to blend in with the dark-colored ocean floor below.

Another kind of defense fishes have is **advertising**. These fishes have sharp quills or poisonous spines. If a predator tries to eat them, the quills or spines will hurt, if not kill, the predator. Of course, the predator might kill the fish in the process, so quills and spines by themselves don't really protect the fish. What protects the fish is the fact that it advertises, or warns, the predators about its spines or quills. These fishes advertise by being brightly colored with bold patterns, which tell predators to stay away.

Notice the lionfish in the picture on the left. Although it is a slow swimmer, it doesn't have many predators brave enough to eat it because they seem to know it is poisonous. Scientists assume predators know this because of the lionfish's bright colors and bold pattern.

A lionfish advertises the fact that it is poisonous.

The best defense for some fishes is to swim in large groups called **schools**. Some schools number in the millions. There is safety in numbers. To a predator, a school of fish can look like a big whale heading in its direction, causing the predator to run for cover. Also, if a predator attacks a school, there are so many fish that the chance of any one particular fish getting eaten is low. It's also probably harder for a predator to chase a school of fish, because when it attacks, the individual fish in the school move in different directions. This probably confuses the predator.

Traveling as a school is a form of defense for these fish.

Some schooling fishes, like anchovies, sardines, and herring, swim like a work of art, with all the fish swimming at the same speed, in the same direction, moving in faultless unity, like a perfectly choreographed dance. They keep the same spacing between each other, swerving, twisting, and even stopping all at once. Although there is no definite leader, schooling fishes stay perfectly together. If a predator swims in the midst, they will scatter, but they rejoin the school when all signs of danger have passed.

# Bouncing Buoyancy

The ocean is a great big place. It is so vast, so enormous, that people have only been able to explore a small part of it. Fishes have a lot of space to cover looking for food, shelter, and mates. They are always on the go. Swimming uses a lot of energy, and thus, all fishes need to rest between flapping their fins and wiggling their bodies from side to side. Have you ever wondered what would

happen to a fish if it stopped swimming? Does the fish begin to sink if it's resting from its hard labor?

God designed fishes to be denser than water. Do you know what that means? **Density** tells you how tightly packed the matter in something is. When something is very dense, it has a lot of "stuff" packed tightly together. For example, if you have a statue made out of plastic and an identical one made out of lead, the lead statue will be heavier. That's because lead is denser than plastic, and therefore the lead statue has more matter in it.

A bobber floats on water because it is less dense than water. A lead weight sinks because it is denser than water.

If something is denser than water, it sinks into the water. For example, a cork is not as dense as water, so it will float on water. But if you put a nickel in the water, it will sink, because the nickel is denser than water. Since a fish is denser than water, it will tend to sink to the ocean floor.

However, God gave fishes a special apparatus to keep them afloat in the water. They have been designed with a **swim bladder**. The swim bladder is like a balloon inside the fish's body. It's filled with gas and can get larger or smaller depending on whether the fish wants to remain higher or lower in the water. The swim bladder, then, makes a fish **buoyant** (boy' ant)! Do you know what that word means? If something is buoyant, it is capable of floating.

If the fish wants to rise in the water, will it make its swim bladder bigger or smaller? Think about it. If you have a balloon wrapped inside a piece of clay and you want the clay to sink in water, do you fill the balloon with more air or less? If you want your balloon and clay to float close to the surface of the water, would you fill the balloon with more or less air? That's right. The more air in the balloon, the higher it will float. In the same way, a fish will make its swim bladder larger if it wants to get closer to the surface of the water, and it will make its swim bladder smaller if it wants to go deeper in the water.

How does a fish regulate the size of its swim bladder? Well, when the muscles around the swim bladder squeeze, the swim bladder gets smaller, and the fish sinks. When the muscles relax, the bladder blows up, and the fish rises. The fish can stay at the same depth by squeezing the muscles just the right amount to stay level. Fishes just know how to do this automatically by instinct. Isn't that amazing?

# Smelly Fishes

Most fishes have a very powerful sense of smell. To understand how a fish smells, you first need to realize that when you smell something, you are actually detecting chemicals in the air. You see, when your mom bakes a nice batch of cookies, some chemicals from the cookies float into the air. If you are somewhere near, some of those chemicals will end up entering your nose when you breathe. Nerve endings in your nose will detect those chemicals and send signals to your brain. Your brain then tells you that you are smelling mom's cookies and that you had better ask her nicely if you can have one.

In order to smell, then, you must breathe air into your nose. You breathe air in through your two nostrils, and you breathe out the same two nostrils. Interestingly, God designed fishes with four nostrils, called **nares**. Two nares take water in, and the other two spit water out. Inside the nares are nerve endings that can detect chemicals in the water. Chemicals travel quickly through the water on currents, allowing fish to smell what is in the water. Sometimes the smells have traveled on currents from miles and miles away to reach the nares of a fish. A fish can often find its next meal, then, by following its nares! It's a good thing God gave fishes such a good sense of smell, for in that huge ocean, it is sometimes difficult to find food unless you are able to "see" far off with your nose.

A sense of smell is especially important for migrating fishes. Consider salmon, for example. Adult salmon live in the ocean. However, each year, they travel many miles up freshwater rivers to find the exact place where they were born. It is thought that salmon find their way back to their breeding ground, across hundreds of miles of open ocean, and through hundreds of miles of streams, simply by using their sense of smell. When they are moving from the streams in which they were bred to the ocean, they smell scents along the way that will help them remember their way back. These scents may include the kind of dirt that washes into the river at certain places, certain kinds of rocks that release their chemicals and smells into the waters, and even certain kinds of fishes that inhabit the rivers and streams along the way.

Following their nares sometimes means that salmon must jump up waterfalls in their struggle to get upriver to the place they were born. This actually presents some danger. Look at the picture on the right. In this picture, a salmon is jumping up a waterfall to get back to where it was born. The bear has been waiting for this and will turn this migrating salmon into a tasty meal. Despite the dangers, however, salmon make this migration, because they have the instinct to do so.

This Alaska brown bear is about to catch a salmon as it jumps up the waterfall in order to get back to the stream in which it was born.

# Do You See What I See?

In the great big ocean, it is important to be able to see where you are going. Most fish have good eyesight. If a fish lives where light penetrates the water, it can usually see colors. This is not the case, however, if a fish lives deep down in the ocean where no light reaches. God gave these kinds of fish enormous eyes that can take in lots of light. Even though they don't see in color, they see extremely well when there only a tiny bit of light.

Usually, the eyes are on opposite sides of a fish's head. This is a special design feature God gave fishes since they can't move their necks. Fishes can actually look out each eye and see things on either side of their head very clearly. Unlike fishes, we are unable to clearly see things located at the sides of our heads; we might know what it is, but we can't see details without turning our neck to look at it with both eyes. The ability to see out of each eye separately is called **monocular** (muh nok' yuh lur) **vision**. One eye might see a shark on the right and send that information to the brain, while the other eye might see a cave to the left and send that information to the brain. Can you guess what a fish would do with that information? It would race to the cave to hide!

# Do You Hear What I Hear?

Have you ever been swimming and tried to talk to someone under the water?  The sound was a bit muffled, but you could hear it loudly.  We usually think of the deep ocean as being a very quiet place.  However, for fishes, this is far from the truth.  Sound travels more than four times faster in water than it does in air, and the ocean is full of many noises and sounds that travel quickly through the water.  Our ears aren't built to hear well under water, so the ocean may seem silent to us. To a fish though, it is full of important noises, like a shark attacking a creature nearby, a school of dolphins jumping above, or lobsters clicking along the sea floor.  All these sounds are easily picked up with the well-designed ears with which God equipped fishes.

# Lateral Lines

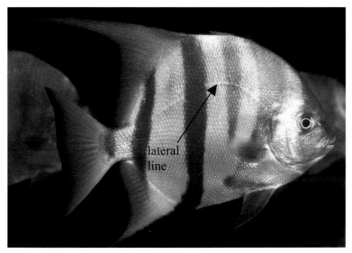

lateral line

The lateral line on this fish is easy to see.  It is not this easy to see on all fish, however.

In addition to hearing and seeing animals in the water, God designed fishes with an extremely sensitive body feature called the **lateral line**.  The lateral line is a line of narrow holes (called **pores**) that runs along the side of the fish's body.  The lateral line usually begins at the gills and ends near the tail.  The pores lead to a chamber filled with groups of tiny sensory hairs.  These hairs can pick up very tiny vibrations that travel through the water.  What does this do for the fish?  Well, by sensing vibrations in the water, a fish can determine what is swimming in the area around it.  If it senses a large fish swimming nearby, it can avoid the potential predator.  If it senses a small fish swimming by, it can capture the fish for a tasty dinner.  If the fish is in a school, it can sense the other fish around it so that it can match its speed to the other fish in order to continue swimming with the school.

# Creation Confirmation

The lateral line is a marvelous work of engineering.  Consider, for example, the groups of sensory hairs that make up the lateral line.  Each group of hairs is called a **neuromast** (nur' uh mast).  There are tens to hundreds of hairs in each group, and they actually convert movement into electricity.  Why do they do that?  Well, the vibrations they are designed to sense involve the movement of water.  However, in order to understand what the lateral line is sensing, the fish's brain must interpret this movement.  The brain gets its information in the form of electrical signals, so the neuromasts actually turn the movement of the water into electrical signals that are then sent to the brain.

Now remember, the lateral line is made up of many neuromasts. Why are there several neuromasts in the lateral line and not just one? It is because several neuromasts located at different points along the lateral line make the system much more sensitive. The lateral line, then, is composed of many sensors, each of which turns vibrations in the water into electrical signals. Those electrical signals are then interpreted by the brain to give the fish the ability to figure out the kind of motion that is going on around it. Some scientists call this sense "touch at a distance," because it allows the fish to "feel" around its environment, sensing motion without ever seeing it. This can be especially useful when the water is murky or when a fish is trying to swim at night.

In fact, scientists have tried to create an artificial version of the lateral line so that robots, which have been made to swim in the water, could have the same kind of "touch at a distance" sense that real fishes have. Of course, the lateral lines created by scientists are crude in comparison to the lateral lines found in fishes. If all the wonders of human science and technology can only produce a crude imitation of the lateral line, it is clear that the lateral line is the work of a Master Designer. It is truly a gift that God gave to fishes, allowing them to "reach out" and sense the world around them.

# Spawning

In most fishes, the female lays eggs that either fall to the bottom of the body of water in which she lives or float with the currents. The males swim about, searching for these eggs. When they are located, the males release a substance called **sperm** on the eggs, which **fertilizes** them. Once this happens, baby fish can begin forming inside the eggs. This process is called **spawning** and is the most common way that fish reproduce. However, God designed other methods for fish to multiply.

Baby sea horses are emerging from this male sea horse's pouch.

Sea horses have probably the most interesting method of reproduction. Sea horses do not lay eggs on the bottom of the ocean; instead, the female places her eggs in a special pouch inside the father. She then departs, never to be involved again. The eggs are kept safe inside the father while they hatch and grow. He is the sole parent and even gives his young food from his body. When the young are ready to venture out into the world, they shoot out from an opening in his pouch, and it looks like the father is actually giving birth.

# Stages of Life

There are five stages of development with most fishes: (1) the **egg stage**, (2) the **larval stage**, (3) the **postlarval stage**, (4) the **juvenile stage**, and (5) the **adult stage**. The first four stages are illustrated on the next page.

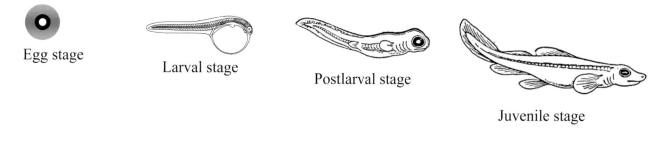

Egg stage

Larval stage

Postlarval stage

Juvenile stage

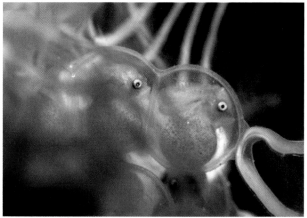

These are anglerfish eggs. You can easily see the developing baby fish inside.

The egg stage, not surprisingly, occurs while the baby fish is growing inside the egg. Most fish eggs develop and hatch within a few days. Sometimes the eggs drift about in the water, while other times, they are attached to a rock or something else at the bottom of an ocean, lake, or river. Other times, of course, the eggs develop inside the parent, as you have already learned. Remember, if the egg develops outside the parent, the fish is said to be oviparous, but if it develops inside one of the parents, the fish is said to be ovoviviparous.

Do you know what caviar is? Caviar is fish eggs. People harvest the eggs by killing the mother fish and removing the eggs from her body before she lays them. Caviar is very expensive because getting the eggs is no easy task. I've never tasted caviar, but it's considered a delicacy (something *really* good to eat) in many countries around the world. In the end, though, it's just fish eggs.

If the fish eggs survive and are not eaten by the many egg-eating creatures in God's creation, they hatch into **larvae** (lar' vee), which is the plural form of **larva** (lar' vuh). At this point, then, the fish are in their larval stage. They float about on the currents, and are therefore considered plankton. Even though a fish in its larval stage is out of the egg, the yolk sac from the egg is still attached to its body. That's where it gets its food while it is in this stage of its development.

These anglerfish are in their larval stage. You can tell by the fact that they still have yolk sacs.

When the fish stops feeding from its yolk sac, it is in the **postlarval stage** of its development.    At this point, it does not have well-developed fins, so it cannot swim well.  As a result, it is tossed to and fro by the currents and is still considered a part of the plankton community.  As the fish's fins begin to develop, it can start to swim on its own.  Before its fins are fully developed, a fish in its postlarval stage is sometimes called a **fry**, which is where we get our term "small fry."  Once its fins are fully developed, the fish can swim against the currents and is in its **juvenile stage**.

Compare the juvenile snapper above to the adult snapper below.  The juvenile looks similar but not identical to the adult.  The juvenile would also be a lot smaller than the adult if the pictures were sized properly.

While a fish is in its juvenile stage, it resembles the adult fish but is still small and immature.  It is important to note that this is really the first stage in which it has much resemblance to its parents.  As the fish matures through its juvenile stage, it develops into an adult.  When it becomes an adult, it is able to spawn and produce new life!

*Tell someone about the different stages of life that most fishes go through.*

# Hermaphrodites

Hermaphra-whats?  **Hermaphrodites** (her maf' ruh dytes) are another example of God's amazing design in nature.  Hermaphrodites are animals that have both male and female characteristics. If someone asked a hermaphrodite animal if it was a boy or girl, the animal would have to say, "Both!" For most fishes, it takes both a male fish and a female fish to reproduce.  Fishes that are hermaphrodites, however, don't have to worry about that.  Any two of them can reproduce.

Some types of fishes are hermaphrodites in another way.  They begin life as males, and as they grow older, they become females.  In a group of these fish, then, all the young fish are males, and all the older fish are females.

# Explore More

Because there are so many different kinds of fishes in the sea, we don't have space in this book to discuss them all.  If you would like to learn about some other fascinating fishes, visit the course website listed in the introduction to this book.  There, you will find links to a lot of information about the various kinds of fishes that God created.  Exploring those links will give you quite a fishy feeling!

# What Do You Remember?

What makes a fish a fish? What fish shape is designed for fast swimming? What did God give fish to help them stay buoyant in the water? How many nostrils does a fish have? Name two defenses that a fish might have. What does "osteichthyes" mean? What does a fish's lateral line do? What is spawning? Name a fish that makes a long journey in order to reproduce. What are the five stages of a typical fish's development?

# Notebook Activities

Label a fish for your notebook. Include the different fins as well as the operculum. In addition, draw a fish for each of the shapes that you learned previously (fusiform, eel-like, depressed, and compressed). The *Zoology 2 Notebooking Journal* provides pages for these activities.

**Younger Students:** Design your own fish using clay or crayons and paper. What is your fish's name? What shape fins will it have? What shape body will it have? What colors will your fish be? Will it be able to use camouflage? How will it protect itself? Will it be a fast swimmer? What will it eat? How will it capture food? Where will it live? Describe its environment. Take a picture of your fish and put it into your notebook. If you are drawing your fish, there is a page in the *Zoology 2 Notebooking Journal* that you can use for this assignment.

**Older Students:** Do some research on the life cycle of salmon. Map out where they begin their lives and the routes they take on their journeys. There is a page in the *Zoology 2 Notebooking Journal* for you to record your findings.

# Ocean Box

There are so many different shapes, sizes and colors of fishes. Use your imagination to create several kinds of fishes that you can put in your ocean box. You can even create a school of fish by shaping tiny fish out of clay and gluing them to the back wall of the box. This will make it look like the fish are all swimming together.

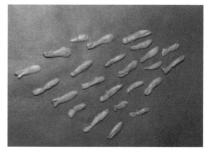

# Optional Experiment

You learned that fishes are ectothermic. This means that the body temperature of a fish changes with the changing temperature of the water. This, of course, has an effect on the fish. Do you think that you can notice the effect that changing water temperature has on a fish? If you don't mind having a pet goldfish at the end of this experiment, let's find out. Think about how a fish might behave in warmer water as compared to colder water. Make a hypothesis about what you might be able to see if you observed a fish in warmer water and then observed it in cooler water.

**You will need:**

♦ A "Scientific Speculation Sheet"
♦ A fish in a small bowl or a glass (A goldfish is best, because it can handle large changes in water temperature. Do not use a tropical fish, as it will not be able to handle large temperature changes.)
♦ A lamp with a bendable neck that can face down into your fish bowl (a 100-watt light bulb)
♦ A large bowl of ice water into which you can place the fish bowl or glass
♦ A thermometer (It should read temperatures above and below room temperature.)

1. Make a hypothesis about how the fish's behavior will change as the water warms up and cools down. Record that hypothesis on your "Scientific Speculation Sheet."
2. Watch the fish in the bowl or glass carefully. In particular, watch its breathing. Notice that the fish opens and closes each operculum to breathe. How many times does it do this each minute? Record this and any other observations you make on your "Scientific Speculation Sheet."
3. Use a thermometer to record the temperature of the water.
4. Now you want to warm up the fish's water. Bend the lamp so that the light bulb points down into the bowl or glass. Make sure that the light bulb stays out of the water, however.

5. Let the light shine on the water for a while, and then use a thermometer to record the temperature of the water. Let the light shine on the water until it is 80° Fahrenheit (27° Celsius).
6. Once the temperature has increased to 80°, observe your fish again. What does your fish do in warm water? Watch its breathing. How many times does it take a breath each minute? Record your observations on your "Scientific Speculation Sheet."
7. Now cool down the water in the bowl or glass by placing it in a larger bowl of water. The water level in the larger bowl should not reach more than halfway up the fish's bowl or glass. Gradually begin to add ice to the water in the larger bowl.

8. Use the thermometer to measure the temperature of the fish's water until it reaches 50° Fahrenheit (10° Celsius).
9. Observe how the fish behaves in the cool water. What does it do? Does its breathing seem to be different in any way? Record your observations on your "Scientific Speculation Sheet."
10. Compare your results to your hypothesis. Were you correct?

Make sure you bring your new pet back to room temperature. What explains the observations you made? The fish's internal body temperature increased in the warmer water and decreased in the cooler water. The warmer the inside of the fish, the more active it can be. Thus, the fish probably breathed more and was more active in the warm water than it was in the cold water. Enjoy your new pet!

# Lesson 7
# Sharks and Rays

Do you know what **cartilage** (kar' tuh lij) is? Feel the end of your nose; bend it back and forth and squeeze it a little. Now feel your ear, especially the top part. Can you bend it this way and that? It is soft and flexible, yet it keeps its shape. That's because it is made of cartilage, not bone. Cartilage gives your nose and ears their shape and structure, but it's lighter and more flexible than bone.

Do you realize that God designed certain fish with skeletons made out of cartilage instead of bone? Not surprisingly, they are called **cartilaginous** (kar' tuh la' juh nus) **fishes**. We'll talk about three groups of cartilaginous fishes in this lesson: **sharks**, **rays**, and **agnathans** (ag na' thunz). Because these fish have skeletons made out of cartilage rather than bone, they are lighter than bony fish, and more flexible, too.

## Sharks and Rays

Sharks and rays are in a class called **Chondrichthyes** (kahn drik' theez). Can you find the Greek word for fish in that word? It's near the end. The first part, "*chondros*," comes from a Greek word meaning "cartilage." Therefore, "Chondrichthyes" means "cartilaginous fish."

There are more differences between bony fish and cartilaginous fish besides the way their skeletons are made. For one, bony fish have scales that lie flat against their skin and are covered with slime for protection against bacteria. Cartilaginous fish, on the other hand, have scales that are more like tiny teeth. In fact, they are called **dermal** (dur' mul) **denticles** (den' tih kuls), not scales. "Dermal" means "skin," and "denticle" refers to teeth. In other words, sharks and rays have teeth in their skin! These "skin teeth" are so deeply embedded in a shark's skin that they are hard to see. Nevertheless, they are tiny, sharp spikes that point in the direction of the tail. Though you can't see them, you can feel them if you run your hand along a shark's skin from its tail toward its head. If you do that, the skin will feel rough and prickly. However, if you run your hand from the head to the tail, it will feel much smoother.

A shark does not have the kind of scales that a bony fish has. Instead, its skin is covered in tiny teeth-like structures called "denticles."

Another difference between bony fish and cartilaginous fish can be seen in their fins. Unlike the thin, almost transparent fins of bony fish, the fins of sharks and rays are thick and rubbery.

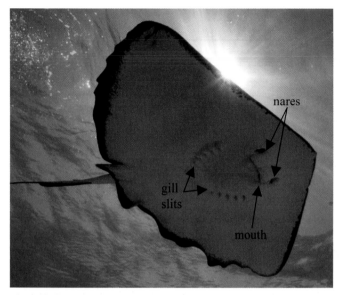

The gill slits, nares, and mouth are on the underside of most rays.

Do you remember what God gave bony fish so they will be buoyant in the water? He gave them a swim bladder. He did not give swim bladders to sharks and rays. Because of this, they sink when they are not actively swimming. This is one reason you will see sharks in aquariums continuously moving, never just floating in the water as fish do. When sharks and rays quit swimming, their density causes them to sink to the bottom of the ocean. This isn't a problem, however, as many sharks and rays are bottom feeders, preferring to wander about on the bottom of the sea in search of food. In fact, most rays are specially designed for bottom feeding, with mouths underneath their depressed bodies.

Another difference between bony and cartilaginous fish is the way God designed their gills. Bony fish have a cover for their gills. Do you remember the name of that cover? It's called the operculum. Cartilaginous fish do not have an operculum to cover their gills. Instead, they have gill slits, which are openings in the skin that lead to the gills.

Do you remember how a bony fish gets water to its gills? It takes in water through its mouth. That water flows over its gills and out its operculum. Because most rays feed and rest on the bottom of the ocean, taking in water through their mouths is not a good idea, because they would probably get a mouthful of sand. Because of this, God gave them breathing holes on top of their bodies called **spiracles** (speer' uh kulz). Rays use their spiracles to take in water that is then sent to the gills. Sharks also have spiracles, but they are usually smaller than the spiracles that rays have. This is because when sharks are swimming around in the water, they can take in water through their mouths, like bony fish. This is

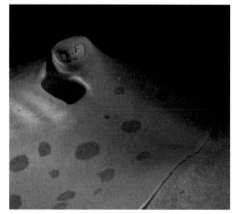

The big hole behind this ray's eye is one of its spiracles.

another reason sharks tend to swim constantly rather than rest. If they rest, the water does not flow through their mouths and go to their gills. As a result, when they rest, they have to rely on their spiracles to send all of the water to their gills.

As you can see, there are many differences between bony fish and cartilaginous fish. Nevertheless, they are all still fish. Do you remember what it is that makes a fish a fish? That's right! If it breathes with gills and has fins, it's a fish.

*Explain some of the differences between bony fish and cartilaginous fish.*

# Rays

Rays look like bats in the water and are actually called **batoids**! God designed them with interesting pectoral fins that are attached to their bodies from back to front, similar to the way a bat's wings attach from its shoulder to its feet. The fins of the ray create a disk shape that can be round, triangular, or diamond-shaped. Rays use these pectoral fins to fly through the water. They are very majestic looking when they swim, like a giant bird gracefully soaring through the ocean.

This stingray was caught in a tide pool when the tide went out. It is fanning its pectoral fins to cover itself with sand for camouflage.

Interestingly, rays tend to be curious, social animals. Though they often live alone, they enjoy swimming together in groups with other rays. Despite the fact that they are able to swim, most rays are bottom dwellers, feeding on benthic animals. In fact, they often flick sand on their backs to camouflage themselves as they lie in the sand at the bottom of the ocean.

To eat, rays use their wide fins to scoop up sediment from the seafloor, looking for clams, oysters, and other benthic animals. They crush their prey with powerful jaws, easily breaking open any shell or exoskeleton the animal has.

# Stingrays

The majority of rays are **stingrays**. Stingrays (along with eagle rays and electric rays) are considered true rays. Near a stingray's tail is a sharp, venom-filled spine used for defense. This sharp spine is like a jagged knife that can whip around and cut a slice in any predator that tries to harm the ray. A stingray can grow to be 8 feet wide, and its jagged spine can sometimes grow to be 14 inches long! Believe it or not, it isn't the cut that hurts the most. It's the venom that the ray releases into the wound. This venom causes immediate and intense pain. Now, that's a powerful weapon.

Have you ever been to an aquarium that had a shallow pool full of stingrays? If so, you might have been allowed to pet them. You may have been afraid to because you heard about their weapon. There is no need to worry, however. Aquarium staffers remove the stingers so that they cannot hurt you. Even in the wild, most rays are meek bottom feeders that rarely use their stingers unless attacked.

Unfortunately, stingrays can hurt an unsuspecting beachgoer who accidentally steps on one that is well camouflaged. As I said, however, rays are not aggressive creatures; they will swim away before they sting if they have the chance. So, if you are wading in the ocean and are afraid of being stung by a ray, just move slowly through the water, shuffling your feet as you go. Any rays in the area will hear you coming and have ample time to get away.

This blue spotted stingray's tail holds a sharp, jagged spine that can slice into a potential predator. As long as you are careful, however, it will not use its spine on you.

# Manta Rays

A manta ray uses its cephalic fins to guide water into its mouth.

Although most rays can be found in shallower waters, submerged in the sand just off the coast, the enormous **manta ray** swims in the open ocean, flying about with its huge mouth wide-open to catch plankton and other debris in the water. The manta ray is considered a filter feeder, using two cephalic (seh fal' ik) fins to guide food into its mouth. It is a docile, harmless creature that doesn't have a stinger. It can grow to enormous proportions – up to 22 feet wide and weighing up to 3,000 pounds. Think about that. Your car probably weighs about 3,000 pounds! Measure the width of the room you are in right now. Is it 22 feet wide or wider? If not, that means a large manta ray wouldn't fit in it!

Sometimes manta rays will actually leap out of the ocean and into the air. Many years ago, sailors were terrified of these rays when they flung their enormous bodies into the air near their boats. They named them "devilfish" because of their fears. So today, manta rays are still called devilfish, though the name is not at all fitting for these wondrous creatures. The truth is, manta rays are among the gentlest animals in God's creation. They are actually so docile that people can swim with them. A favorite activity of scuba divers is to hitch a ride on a manta ray as it glides in slow motion through the open seas.

With its huge mouth positioned on the front of its body, opened wide to draw in floating plankton, the manta ray generally stays near the surface of the ocean, continuously flapping its wing-like fins. When feeding, it often swims in a complete circle. The circular motion causes the plankton to get caught in the whirling currents it creates, concentrating the creatures in one spot so the manta ray can get its fill. Because zooplankton tend to rise to the surface of the ocean at night, guess when mantas feed? At night, of course! Do you remember that plankton are often found close to the shore, where the phytoplankton is especially thick? This means mantas can also be found close to shore, especially in the evenings. In some places, such as Hawaii, a moonless night will attract billions of plankton to the lights of the hotels on the beach. The plankton congregate there, and so do the mantas! People enjoy watching the spectacle of the mantas feeding.

Manta rays frequently visit the **cleaning stations** found in coral reefs. A cleaning station is where certain fish (and sometimes shrimp) clean the teeth or skin of other fish. They do this by simply eating whatever has accumulated on the teeth or bodies of the fish that have come to be cleaned. Enormous manta rays line up here, one behind the other, each waiting its turn. When a manta's turn comes, the cleaner fish eat all of the parasites that have clung to the manta. This feeds the cleaner fish and keeps the manta free from disease. In other words, the mantas and the cleaner fish were made for each other.

# Electric Rays

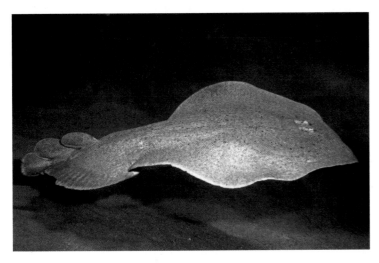
This electric ray, known as a numbfish, can deliver a powerful shock!

One interesting ray is the electric ray. There are about forty different species of electric rays in the ocean. These rays have battery-like electric organs on both sides of their heads, which can produce a powerful electric shock of up to 240 volts! Most electric rays produce a smaller 20 to 40 volt shock, however. While this is enough to cause a person some pain, it is typically not enough to cause injury. Electric rays use this shock to stun prey and for self-defense.

Most electric rays have rounded bodies and live near the shore in sea grass beds. They are considered very lethargic, meaning they don't tend to move much at all, even when disturbed.

# Eagle Rays

Eagle rays have long whip-like tails. Although their long tail is equipped with a stinger, like the stingray, they are not classified as stingrays. Eagle rays are also recognized by their especially pointy snouts and diamond-shaped bodies. They swim differently from stingrays, flapping their fins like wings. Stingrays swim by wriggling the outer edges of their fins.

Though eagle rays are bottom feeders, they don't dwell on the sandy floor as stingrays do. In fact, these beautiful animals often swim on the surface of the ocean, sometimes leaping out of the water in amazing displays. They prefer to swim in groups, which look like clusters of kites floating just under the surface of the water.

Eagle rays "fly" through the water by "flapping" their pectoral fins.

# Sawfish

One ray that has not undergone much study is the sawfish. This ray has a long body, much like a shark. However, it has its mouth, nares, and gill slits on the underside of its body, just like a ray. Also, its pectoral fins are enlarged, like those of a ray. Thus, even though it has some body characteristics in common with a shark, it is mostly a ray, so it is classified with the rays.

Sawfish are strange-looking rays.

One feature that separates a sawfish from all the other rays is its long, saw-like snout. On either side of its long snout are little teeth like those of a saw. These are not like the teeth you find in a shark's mouth. Instead, they are actually large denticles. In addition, unlike a shark's teeth, these teeth aren't replaced if they are broken off.

Why does the sawfish have a "saw" to begin with? Well, it tends to rest at the bottom of the ocean, and its "saw" is covered with keen sensors that can detect the movements and even the *heartbeats* of creatures that are buried in the sand. It can then use its saw to dig up its prey. In addition, if a hapless fish swims over a sawfish, the sawfish will jump into action, slashing the creature above with its saw. This stuns or kills the fish, allowing the sawfish to enjoy a nice meal. The saw can also be used for defense.

These students are examining the snout of a sawfish. Notice that there are "teeth" missing, because they are not replaced if they are broken off.

The longest sawfish is over 20 feet long. These rays can be found both off the coast and in rivers that empty into the ocean.

# Skates

Notice the thorns on this Melbourne skate's tail as well as the extra pair of fins near the base of the tail.

Skates are similar to stingrays, but they tend to be smaller with shorter, thicker tails. There is also another major difference: skates don't have stingers. Instead, their tails are thicker with small thorny growths along the edges. Since skates don't have stinging spines, they depend on these special thorns to deter predators. At the base of the tail is a set of extra fins as well.

Skates and rays also differ from each other in the way they give birth. Most rays are ovoviviparous. Do you remember what that means? It means the eggs develop inside the mother's body. Skates, on the other hand, lay their eggs. This means they are oviparous.

# Sharks

Very few things seem as frightening as a shark lurking near the place where you are swimming. And, well, probably nothing is. However, many sharks are bottom dwellers that scavenge the ocean floor, feeding on clams and crabs. In addition, the largest fish on earth, the whale shark, which can be the size of a city bus, is a docile, harmless creature that wouldn't hurt a flea. That is, perhaps, except for a water flea, which is a kind of plankton. That's because the biggest fish in the sea feeds on the smallest organisms in the sea, plankton.

Although many sharks are relatively harmless to people, great white sharks like this one have been known to attack and kill people.

Sharks are fascinating creatures that haunt the seas. Their bodies are streamlined in a perfect fusiform shape for fast swimming. They are so fast, in fact, that sharks are the only creatures able to catch tuna or marlin swimming in the open sea.

Though we imagine fearsome stories when we think about sharks, of the more than 250 species of sharks, only about thirty are known to have attacked people or boats. The most common attackers are the great white sharks, tiger sharks, bull sharks, and hammerhead sharks. However, though billions of people go to the beaches all around the world every year, only about 75 shark attacks occur each year. To put this in perspective, Sea World posted this information: "Worldwide, 3 people died due to shark attacks; in comparison, 150 people died from coconut strikes, 200 people died due to elephants, and 2 million people died from not washing their hands and contracting bacterial diseases." So, it appears that swimming in the ocean is much safer than, well, eating — at least with unwashed hands.

# Shark Teeth

The skin surrounding this shark's mouth has been removed so that you can really see the teeth. Notice that they are lined up in rows.

Perhaps the most interesting feature of a shark is its teeth. As you may know, sharks have several rows of teeth, each row hidden behind the row in front. A shark can have more than a hundred teeth in the front row. When a front row tooth breaks, a new one from the row behind it moves up to take its place. The teeth from the other rows move up as well, and then a new tooth grows in the back row. Why do sharks have rows of teeth? They eat so violently that they often break a few teeth. However, they cannot wait for their front-row teeth to grow back, so they need "backup teeth" that can replace their front-row teeth very quickly. Then, the new teeth can grow slowly in the back row, because they won't be used for a while. A shark might have up to three thousand teeth inside its head at any one time. Over the course of its lifetime, it may lose and regrow as many as *thirty thousand teeth!*

With all these teeth, you would think sharks could chew, but they can't. So they bite their prey and then jerk it around in order to yank off a chunk to swallow. I know that's not a very nice thought, but it's the way a shark eats. The chunks of food that a shark swallows end up in its stomach, where they are digested. This is a slow process, however, so a meal might take several days to digest. This is why sharks don't eat every day.

Shark teeth come in many shapes and sizes. Some sharks have sharp, pointy teeth, while bottom dwelling sharks have cone-shaped teeth for crushing shells. Because there are so many different kinds of sharks, and because each has its own kind of teeth, many people enjoy collecting shark teeth. Shark teeth collectors can estimate how big a shark was by measuring the shark tooth. First, they measure the length of the tooth in inches. Every inch of tooth equals 10 feet of shark length; so if a shark tooth is 2 inches long, the tooth came from a shark that was 20 feet long.

**Try This!**

The picture on the right is of real shark teeth, shown in their actual size. The one in the middle belonged to a **megalodon** (meg' uh luh don'), an enormous shark that is now extinct. With a ruler, measure these teeth and see if you can estimate how big the sharks were!

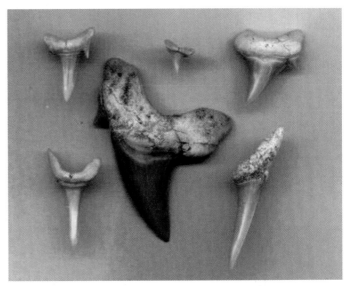

# Shark Sense

God designed sharks with many amazing features that help them find their prey. In certain conditions, they can sense prey miles and miles away. Let's discuss how they can do that. Look at the shark image below. Can you see the tiny holes all over the shark's snout, especially between the eye and the tip of the snout? Those holes are not used for smelling; instead they are nerve receptors called the **ampullae** (amp yoo' lee) **of Lorenzini** (lor en zee' nee). What do they do? They actually sense electricity in the water. But what gives off electricity? Why, every animal in the water gives off electricity!

The tiny holes that cover this shark's snout are its ampullae of Lorenzini. They are easiest to see between the eye and the tip of the snout.

Every time an animal's heart beats or it moves, tiny currents of electricity are produced. These faint electrical currents produce signals that travel easily through water. Sharks and rays can sense these faint electrical signals. In fact, sharks use this sense even more than sight when catching prey. When their prey is close at hand, for example, they roll their eyes back into their head, using only their ampullae of Lorenzini to finish the kill.

Interestingly, when a feeding frenzy occurs somewhere in the ocean (like when a school of tuna finds a school of herring and begins feasting) there is a lot of activity. Because of all the activity, there are a lot of electrical signals being produced in the water. It's no wonder that sharks often show up at these feeding frenzies!

Many sharks have good eyesight, and most have the ability to smell exceptionally well. In fact, two-thirds of a typical shark's brain is devoted to giving it its sense of smell. This allows sharks to sense smells that are miles and miles away. Sharks are especially sensitive to the smell of blood and will rush toward it, following the tiny blood particles that have been carried by the currents from a distant bleeding animal. Like all fish, sharks also have a lateral line, which detects vibrations in the water. So, with good vision, the

Great white sharks have been observed making sudden leaps out of the water to snatch surprised seals swimming in the water.

ability to sense tiny electrical signals, the ability to smell blood from miles away, and their lateral line, sharks have an advantage over their unsuspecting victims.

But *finding* prey is only half the story. Sharks must also be able to catch their prey. Here again, sharks have the advantage. Sharks possess streamlined, bullet-shaped bodies that slip through the water at top speed. A shark's tail fin usually has a larger lobe on top than on bottom, and it can also be seen projecting out of the water along with its dorsal fin when the shark is swimming close to

You can see both the dorsal fin and the tail on the shark (top), but only the dorsal fin on the dolphin (bottom).

the surface. In fact, that's one way to tell the difference between a dolphin skimming the surface of the water and a shark doing the same thing: a dolphin shows only its dorsal fin, while the shark often shows both its dorsal fin and the upper lobe of its tail. So, if you are at the beach and see a fin moving along the top of the water, it's probably a dolphin. If you see a fin with a second fin projecting up right behind it, it's probably a shark. If you find yourself in the water with a shark close by, which will probably never happen, follow the instructions at the end of this lesson so you can escape unharmed.

Different sharks don't prey on the same animals. Though they'll eat most anything if food is scarce, each kind of shark has a preference for certain creatures. Hammerhead sharks prefer to eat stingrays, for example, while the dogfish shark feeds almost exclusively on crabs and lobsters. The tiger shark is a fierce predator of sea turtles, while the whale shark and basking shark gulp down huge amounts of plankton each day.

# Creation Confirmation

A shark's ampullae of Lorenzini are yet another example of how incredible God's creations are. Not only can a shark detect the presence of prey by using its ampullae of Lorenzini, but the shark's brain can even interpret the signals so as to give the shark the precise location of the potential meal. This combination of the ampullae of Lorenzini and the shark's brain is so amazing that a shark could theoretically detect the precise location of a 9-volt battery from over 1,000 miles away! To this day, human science and technology have not come close to producing such a marvel. The very fact that the shark naturally has such an amazing system tells us that the shark could not be the result of chance. It was clearly designed by an amazingly intelligent Designer!

# Shark Pups

Most sharks are ovoviviparous. As you know, this means shark eggs usually develop and hatch inside the female's body. As a result, sharks are born fully developed and are able to swim and catch prey as soon as they are born. A female shark is pregnant for a long time. It can take almost two years for a shark to develop and emerge from its mother.

Because a shark is a shark, and preying upon animals is what most sharks just naturally do, what I'm about to tell you shouldn't really surprise you, though it may sadden you. You see, a few of these baby predators growing inside their mother won't make it to their first birthdays. Why is this? Well, in many sharks, the pups attack and eat the other pups growing inside their mother. That is sibling rivalry at its very worst. These predators start early…very early, indeed.

Not all sharks give birth to live young. Some sharks will lay egg cases, which are often called "mermaids' purses." These come in several shapes and sizes, depending on the shark that laid them. The shark will develop outside of the mother's body, inside the egg case. When the shark is fully developed, the egg case splits, and a baby shark swims out. These egg cases will often wash up on beaches after the sharks hatch from them. Shark egg cases are generally rectangular, sometimes with stringy tendrils attached to the four corners. Other sharks lay spiral-shaped egg cases.

This is a shark egg case.

*Tell someone all that you can remember about shark teeth, shark senses, and shark pups.*

# Shark Orders

There are several different kinds of sharks that swim in the oceans. To give you an idea of just how many types of sharks there are, I want to tell you about the major orders into which scientists classify sharks.

Let's start with order **Squatiniformes** (skwah' tin ih for' meez), which contains **angelsharks** and **monkfish**. A shark in this order has a flat body with a mouth on the front of its head and no anal fin. They are flat like rays, and could even be mistaken for rays. The pectoral fins of a shark in this order resemble the wide fins of a ray, and the eyes are on top of the head. Because they are bottom feeders, they also use spiracles on their head to bring water to their gills. They generally lie still on the bottom of the ocean floor, and when prey moves

This angelshark can be mistaken for a ray.

past, they pounce out of hiding and snap it up. Their prey consists of shrimp, clams, and other benthic animals. These sharks can grow up to 6 feet long.

Dogfish sharks are the most common type of shark.

The next order, **Squaliformes** (skwah' lih for' meez), contains deep-water sharks that are rounded in the middle and have short snouts. They have two dorsal fins and no anal fins. The second dorsal fin is generally smaller than the first. The dogfish shark is an excellent example of the sharks in this order. They are called dogfish because they form packs (schools) for hunting and traveling. The spiny dogfish is the most common shark in the sea. It is small, about 2½ feet long, and forms groups of hundreds or thousands of individuals of the same sex and size. It is called a spiny dogfish because the front of each dorsal fin has a sharp, pointed spine.

Order **Hexanchiformes** (hex' an kuh for' meez) contains the sharks that have "extra" gill slits. All other sharks in the ocean have five gill slits, but members of this order have either six or seven. They each have an anal fin and a single, limp dorsal fin. They spend most of their time in very deep ocean water. Cow sharks and frilled sharks are both members of this order.

This frilled shark looks more like an eel, but because it has gill slits and not an operculum, you know it is a shark.

This bullhead shark is designed to eat benthic animals.

Order **Heterodontiformes** (het' uh roh don tih for' meez) contains the bullhead sharks and hornsharks. These sharks typically have large heads and pig-like snouts. They live in coral reefs, eating crabs, shrimp, sea urchins, clams, and other benthic creatures. They have two spiny dorsal fins and an anal fin. They grow to be about 5 feet long and lay interesting-looking, spiral-shaped egg cases.

Order **Orectolobiformes** (oh rek' toh lob ih for' meez) contains sharks like the whale shark and nurse shark. These sharks have their mouths on the fronts of their heads, rather than tucked below. They also have barbels on the nostrils that make them look a little like catfish. Some sharks in this order are quite small, but this order also contains the biggest fish in the world, the whale shark, which grows up to 45 feet long! The whale shark has a huge mouth that can open 4 feet wide. It usually has pale yellow stripes and dots on its four-inch thick, gray skin. This giant shark is no danger to people, as it is a filter feeder, opening its large mouth to pull in huge amounts of plankton. As the water passes over its gills, bristly structures called **gill rakers** catch the plankton that are in the water. Anything caught in the gill rakers gets eaten.

The scuba diver in this picture gives you some idea of how big this whale shark is.

Sharks in order **Lamniformes** (lam' nih for' meez) grow fairly large. Although this order contains many sharks, such as the mako shark**,** sand tiger shark**,** and megamouth shark, the best-known shark in this order is the **great white shark** (see pictures on pages 109 and 112). It is the best-known shark because it is often responsible for attacks on people. The majority of shark attacks are on surfers. This is probably because, while a surfer is paddling out to "catch a wave," he or she looks much like a seal or sea turtle from below. People often survive shark attacks, however. This is most likely because, after taking one bite, the shark realizes that the person it is biting is not the seal or sea turtle that it wanted.

The great white's metabolism is so slow that once it has eaten, it can go months before eating again. Great whites grow to be 22 feet long. Therefore, they are too large to spend a lot of time in the shallow waters near the shore. Although several dangerous sharks are in order Lamniformes, so also is the huge basking shark, which, like the whale shark, feeds only on plankton.

Sharks in order **Carcharhiniformes** (kar' char hin ih for' meez) are called ground sharks because they tend to wander around close to the seafloor in huge numbers. The members of this order include blue sharks, tiger sharks, reef sharks, and leopard sharks. Many sharks in this group are known to attack humans. They each have five gill slits, an anal fin, two spineless dorsal fins, and a mouth that is behind the eyes. They range in size from small to large.

**Hammerhead sharks** are probably the most well-known sharks in this order. These strange-looking sharks have flattened heads, and their eyes are on the outer edges of the head. This strange head shape has a purpose. It allows hammerhead sharks

The strangely-shaped head tells you this is a hammerhead shark.

to make very tight turns in the water, and it also increases their abilities to sense prey.

*Tell someone all that you remember about the different kinds of sharks in the world.*

# Avoiding Shark Bites

If you are at the beach, there are certain precautions you can take to be less noticeable to sharks. The first step you can take is not to carry dead fish when you are in the water. That may sound silly, but people have done it. The dead fish can attract sharks. Also, sharks hunt early in the morning or early in the evening. They don't like to hunt during the heat of the day. So spend your time in the water during the hottest part of the day, around noon and in the early afternoon. Stay out of water that is dark and murky. Don't wear bright colors that are contrasting because sharks can be attracted to bright colors, especially if they are contrasting.

Don't swim near places where there is a steep dropoff. Sharks like to stay in those steep dropoffs so that they can swim to great depths. Never provoke a shark you see in the water by playing with it or poking it. If you happen to see a shark, stay calm and don't make rapid movements. Swim calmly and in rhythm toward the land or boat. Keep the shark always in sight. Look directly at it, because sharks tend to attack when they aren't noticed. If all else fails, fight the shark with your fists and legs, kicking it in the head and nostrils. People have successfully fought off a shark in this way.

# Jawless Fish

Deep down in the darkest depths of the ocean lives a fierce hunter — a frightening creature. It is one of the few creatures that can easily kill the great white shark or the killer whale. Unaware of its presence, the victim wanders about near the seafloor, when suddenly this dark, slithering, snakelike fish springs out from hiding and burrows its wide open mouth into the side of the shark. Usually, it's not alone, but accompanied by a team. They cut their victim by rubbing their teeth back and forth against skin until they break the flesh. They then begin sucking the life out of the helpless victim. These dark, slimy killers are called lampreys.

Lampreys are placed in class **Agnatha** (ag nah' thuh), which contains a special kind of cartilaginous fish. The Greek word "*gnatha*" means "jaw," and when you place an "a" in front of a word, it means "without." So "Agnatha" means "without jaw." So guess what these fish do not have? They don't have a jaw. They have mouths, but they don't have the ability to open and close them. Open your mouth. Now imagine if you couldn't close it ever again. How would you eat? You couldn't chew because you couldn't close your mouth. This is the life of an agnathan. Its mouth is always open. It does have a muscle that keeps its throat closed when it is swimming, but it can never close its mouth. So how does it eat? It suctions and slurps up its food, much the way you eat a popsicle — except try it without closing your mouth to swallow.

Although lampreys are definitely fish, they look more like snakes. They do have fins and gills, however, which make them fish. Their always-open mouths are filled with horny teeth, which help break through the flesh of their victim. The helpless fish will swim around with the lamprey hanging out of its body until the parasitic lamprey has sucked out so much of its blood that it can no longer live. Lampreys sometimes work in schools, as fish often have several lampreys attached.

gill slits

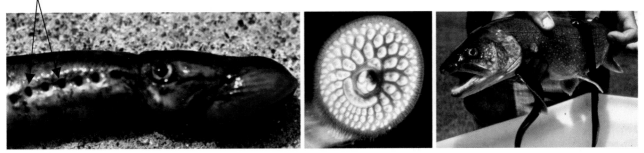

Side View of a Lamprey          Lamprey Mouth          Lampreys Feeding on a Fish

Lampreys have interesting lives. Although they are born and grow up in freshwater, they eventually leave the freshwater and live their adult lives in the ocean. However, they do come back to freshwater to reproduce. When an animal is born in freshwater but lives its adult life in saltwater, we say that it is anadromous (an ad' roh mus). Though most people consider lampreys a nuisance today, believe it or not, during the Middle Ages, lampreys were caught for food. It is said that King Henry I of England died from indulging on too many lampreys.

Another member of this class, the hagfish, looks even less like a fish than does a lamprey. It's practically a worm — a giant earthworm. In fact, though fish are supposed to have backbones, this one does not. Its skeleton mostly consists of its skull, which is made out of cartilage. Yet, it has no other place to fit into the animal world, so, because of its fins and gills, we call it a fish. This creature doesn't eat with suction. Instead, it uses its "tongue" teeth. Its tongue has teeth to nip pieces of flesh from its victim. It doesn't have to hurry though, because its victim is already dead; the hagfish is a scavenger. Its teeth pinch together to lock onto its food, helping it tear into the flesh of dead and dying fish that have sunk to the bottom of the deep, cold sea, where it dwells. Often, a hagfish will actually bore into the dead fish that it is eating, "cleaning out" the insides of the dead fish.

Typically, hagfish are only seen by humans when nets pulled up after sweeping the seafloor. Every fish, even the dead ones at the bottom of the sea, are brought up into the boat by the net. In some of those dead fish, hagfish are found eating. These putrid fish are dumped onto the ship deck with the hagfish protruding from their bodies. When the hagfish are removed from the fish, they do a very interesting trick that repulses fishermen. When frightened, a small hagfish can continuously vomit up such enormous volumes of slime that it will completely fill a 2-gallon bucket. That's a lot of slime, and a seemingly impossible feat for such a small creature. The reason such a small fish can produce more slime than it seems its body can hold is because the slime comes out in fibrous strings that quickly swell up to several times their original size when they are released from the hagfish's body.

With no backbone, the hagfish can perform an amazing feat of tying itself into a knot and slipping out of that knot. Why does it do this? Well, there are two reasons. First, hagfish produce a slime covering that deters predators. This slime covering can pick up a lot of nasty things, especially when the hagfish is wriggling around inside a dead fish. When a hagfish ties itself into a knot and then slips out, the slime is rubbed clean. The second reason is for when it is eating. Sometimes, in order to tear off a bit of food from a dead fish, the hagfish really needs to pull hard. Tying itself in a knot gives the hagfish the leverage it needs to do this.

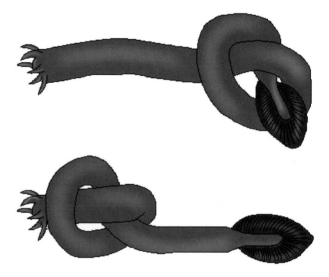

A hagfish can tie itself in a knot and then slip out of the knot. This removes the outer parts of its slime covering.

Note in the drawing above that you can't see the hagfish's eyes. That's because they are small and not very useful. The hagfish uses mostly its senses of smell and touch to find food.

# What Do You Remember?

What does "Chondrichthyes" mean?  What are the scales of sharks and rays like?  How are the fins of bony fish and sharks different?  Why do sharks and rays sink when they are not swimming?  What is the difference between a manta ray and a stingray?  How can you tell the size of a shark by its teeth?  How do the ampullae of Lorenzini help a shark?  If you are swimming in the ocean and see a shark, what should you do?  What does "Agnatha" mean?  What does "anadromous" mean?  What are some differences between the way a hagfish feeds and the way a lamprey feeds?  What do almost all fishes have that the hagfish doesn't have?

# Notebook Activities

After recording some of the interesting facts you learned, draw or find photos of sharks, rays, lampreys, and hagfish. Put them in your notebook and add some facts about each under each picture. There are pages for this activity in the *Zoology 2 Notebooking Journal.*

**Older Students:** Prepare an informative speech, using your own words, stories, illustrations, and possibly some humor to explain what one should do to avoid shark bites. There is a page for this assignment in the *Zoology 2 Notebooking Journal.*

# Ocean Box

Make a shark and a ray and place each in your ocean box.  You can also fashion a lamprey out of black clay and place it so that it hangs off one of the fish you made last week.

# Experiment

Do you remember what the ampullae of Lorenzini do for a shark?  They allow it to detect the electrical signals produced by the movement and even the heartbeat of underwater animals.  This helps sharks find their prey.  What if the shark lived in freshwater instead of saltwater?  Does electricity travel as well through freshwater as it does saltwater?  Let's find out by doing the next experiment.

**You will need:**
♦  A "Scientific Speculation Sheet"
♦  A large battery (See picture on the next page.)
♦  Three pieces of insulated electrical wire
♦  Two metal nails (They must conduct electricity.)
♦  Electrical tape
♦  A glass of distilled water (You can buy this at any large supermarket.  Make sure it is distilled water, not mineral water.)

♦ A small light-bulb holder with a light bulb (These can be found in hobby stores. Look at the picture below to see what I mean.)

♦ Salt

**Procedure:**

1. Write a hypothesis on your "Scientific Speculation Sheet" about whether freshwater or saltwater will conduct electricity better.

2. Strip the insulation off the ends of the wires so that there is bare metal wire exposed.

3. Attach the wires as shown in the image on the right. One needs to be attached to the negative side of the battery. The next needs to be attached to the positive side. One of those two wires (it doesn't matter which) needs to have its other end attached to one side of the light-bulb holder. The last wire needs to be attached to the other side of the light-bulb holder.

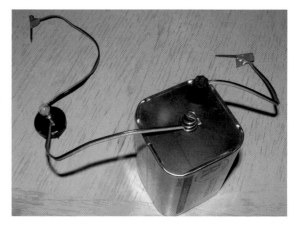

4. Each of the free ends you have left needs to be wrapped around a nail, and then that connection should be secured with electrical tape.

5. **Being sure to hold the wires only where there is insulation**, touch the nails together. This completes the electrical circuit and allows electricity to flow. If you have connected everything right, the light bulb should turn on. If it does not, check your connections and make sure that metal is touching metal at each of them.

6. Pull the nails apart. The light bulb will go out.

7. Fill the glass with distilled water.

8. Once again, **being sure to hold the wires only where there is insulation**, place both tips of the nails into the water. What happens to the light bulb? What does that tell you about how freshwater conducts electricity?

9. Place two teaspoons of salt into the water and stir.

10. Once again, **being sure to hold the wires only where there is insulation**, place the two tips of the nail into the water. What happens? What does that tell you about how saltwater conducts electricity?

Why did the light bulb not light when you used freshwater but lit well when you used saltwater? It's because saltwater conducts electricity better than freshwater. Salt is actually made of positive and negative particles called ions. When the nails were put in saltwater, the positive ions moved to the nail connected to the negative side of the battery, and the negative ions went the opposite way. That allowed electricity to be conducted.

# Lesson 8
# Crustaceans

Imagine this scene: Deep in the ocean, hidden from the eyes of mankind, two sea snails meet in a clash that will end the life of one and prolong the life of the other. A giant **conch** (kongk) devours a whelk. While the conch slowly but methodically consumes the smaller whelk, the smells of the dying creature travel through deep sea currents, alerting a few other creatures who scurry quickly over to the scene. These little creatures wait impatiently while the whelk is being eaten, occasionally jumping on the dying whelk trying to crawl inside its shell. Dozens of little creatures wrestle with one another, hoping to be the first one to get inside the shell. In the end, whoever wins the scuffle

Hermit crabs like this one make their homes in empty shells left behind by animals such as snails.

wins the shell, and whoever wins the shell wins a nice new home. These little scuffling creatures are none other than God's intelligent little hermit crabs. We will study hermit crabs and other creatures like them in this lesson. The giant conch and whelk are sea snails that we'll explore in the next lesson.

Have you ever walked through the meat section of your grocery store and noticed a tank full of lobsters? There they are, with their claws clamped together, waiting to become someone's meal. Look closely and you will see that the entire lobster is covered by a hard shell that protects it like a suit of armor. This shell can be thought of as a crust. Perhaps that's why lobsters (and animals like them) are called **crustaceans** (kruh stay' shunz). Can you name any other animals that are crustaceans? Examples include **shrimps**, **crabs**, **krill**, and the now extinct **trilobites** (try' luh byts).

Most crustaceans spend a good part of their lives crawling along the bottoms of streams, rivers, and the ocean, occasionally coming on land. Do you remember what we call animals that move along the ground under the water? We call them "benthic creatures." They don't swim about as nekton do. Though lobsters and shrimp can swim a bit, their primary mode of travel is to walk along the bottom of the body of water in which they live.

Most kinds of creatures that God created have freshwater and saltwater counterparts. This means that if there is a kind of animal that lives in the sea, a similar kind of animal probably lives in freshwater as well. There are saltwater and freshwater shrimps, crabs, and lobsters. Freshwater crabs and shrimps can even be purchased at pet stores for your aquarium. In fact, have you ever kept sea monkeys as pets? They aren't monkeys. They are brine shrimp, found wild in saltwater lakes.

Crustaceans range in size from large to very small. The North Atlantic lobster, for example, can weigh over 40 pounds, and the Japanese spider crab has legs that can be up to 12 feet long. The water flea, however, is a tiny crustacean that can be less than a millimeter (0.04 inches) in length.

Each place the legs can bend is a joint in this crab's armor.

Crustaceans are members of animal phylum **Arthropoda** (ar thrah' poh duh). Insects are in this phylum as well. "*Arthron*" is the Greek word for "joint," and "*podos*" is the Greek word for "foot." Members of this phylum, often called **arthropods**, have jointed feet and legs. Look at the crab in the picture. Its legs have joints in them where they can bend. This is important, because the "crust" or "armor" that covers it is hard. Without joints, it would not bend. If you look at an insect up close, you'll notice the same kinds of joints in its legs. That's because insects have armor like the crab. Let's discuss this armor.

# Exoskeleton

When you see an exit sign inside a building, what is it telling you? It's showing you the way to get outside, isn't it? Well, the armor that covers a crustacean is called an **exoskeleton** (ek' soh skel' ih tuhn). "Ex" refers to "outside," so an exoskeleton is a skeleton that's on the outside of the body. You and I have our skeletons (our bones) on the insides of our bodies. Crustaceans have no bones inside their bodies, only squishy flesh. Just as the bones in our body hold us up so that we are not just a squishy glob of flesh, a crustacean's exoskeleton holds it up, but it does so from the outside of the body. Also, just as knights used to wear armor over their entire bodies for protection during wars, a crustacean's exoskeleton gives it protection.

Of what is that exoskeleton? It is made of a tough but flexible chemical called **chitin** (kye' tin). In addition, if the crustacean needs a very hard exoskeleton for a lot of protection, the chitin will be hardened with another substance called **calcium** (kal' see uhm) **carbonate** (kar' buh nayt). Believe it or not, you have probably held calcium carbonate in your hand, because chalk is made mostly of calcium carbonate. Some crustaceans, like lobsters and most crabs, have very hard exoskeletons, so they use a lot of calcium carbonate. Other crustaceans, such as shrimps and krill, have fairly thin, more flexible exoskeletons. Because their exoskeletons are not nearly as hard, they don't need or use much calcium carbonate.

Though an exoskeleton is a great covering for a crustacean, it does have one problem: it can't grow larger. So, in order for a crustacean to grow bigger, it must take off its exoskeleton. But how can it do that? Well, it **molts**. To molt, an arthropod produces chemicals that eat away at its exoskeleton. This weakens the exoskeleton. The arthropod then takes in water so that it begins to swell. At the same time, it starts growing a new, flexible exoskeleton under the old, weakened one. When the old exoskeleton gets weak enough, the arthropod breaks out of it. Now because it swells up due to all of the water that it takes in, the arthropod is actually much bigger than it should be. That's good, though. You see, once it has swelled up as much as it can, the new exoskeleton hardens, and the arthropod gets rid of all the excess water. This makes the arthropod inside the exoskeleton smaller, but the exoskeleton stays the same size. That way, the exoskeleton is bigger than the arthropod, so the animal has room to grow before it has to molt again. After the molting is done, the little guy often eats the old exoskeleton. Yeah, that's pretty gross, but the old exoskeleton is filled with a lot of nutrients that God doesn't want wasted.

A newborn crustacean will molt several times a year until it's rather large. Once it's bigger, it molts only once a year. Very old crustaceans, such as lobsters, which can live to be a hundred years old, will molt just once every few years.

## Crustacean Anatomy

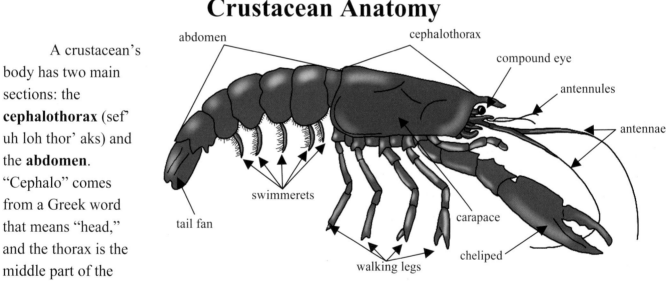

A crustacean's body has two main sections: the **cephalothorax** (sef' uh loh thor' aks) and the **abdomen**. "Cephalo" comes from a Greek word that means "head," and the thorax is the middle part of the body. If you remember your study of insects in *Zoology 1*, you will recall that the head and thorax are two different body parts in an insect. In a crustacean, those two body parts are fused to make one body part, the cephalothorax. The part of the exoskeleton that covers the cephalothorax is called the **carapace**. Have you heard that word before? A turtle's shell is also called a carapace. In science, the same words are used to define similar things. That's why I keep teaching you the meaning of words. If you know word meanings, science will be easier for you.

People who do not study much science often call the abdomen the "tail" of the crustacean. Restaurants, for example, offer lobster tail as a meal. However, when you eat lobster tail, you are not eating the tail of a lobster. You are eating the muscles in its abdomen.

# Head Features

Many crustaceans have two sets of antennae.  The two in the longer set are called **antennae**, and the two in the smaller set are called **antennules**.  The antennules help the crustacean keep its balance and allow it to touch and taste things.  The antennae are much more sensitive, giving the crustacean strong senses of touch and taste.  In some crustaceans, the antennae are longer than the rest of the body!  Please note, however, that not all crustaceans have antennae or antennules.  Crabs are definitely crustaceans, but they have neither.  They get their senses of touch and taste from sensory units on the ends of their walking legs.  Yes, crabs actually step on their food to taste it!

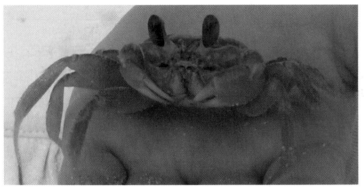

It is easy to see that this crab's eyes are mounted on stalks.  Note that the crab does not have antennae or antennules.

Crustaceans have compound eyes.  You should remember what compound eyes are from your study of insects in *Zoology 1*.  Compound eyes are big eyes made up of many tiny lenses, each of which gives a different image to the brain.  This means that a crustacean sees a mosaic of many images.  While this results in eyesight that isn't as sharp as human eyesight, the combination of images enables the crustacean to detect movement especially well.  Many crustaceans have their compound eyes on top of stalks that can move around.  This is great for a crustacean, because crustaceans don't have necks.  When we want to see something behind us, we just turn our necks so that we can see what's there.  A crustacean that has its eyes on stalks can simply turn its eyes in the other direction so that it can see what's coming up from behind.

Imagine that, in addition to your upper jaw and your lower jaw, you also had little fingers that came out of your mouth to hold your food while you ate it.  That's how God made crustaceans.  Every crustacean has three sets of mouthparts: **mandibles**, **maxillae** (mak sil' ay), and **maxillipeds** (mak sil' ih pedz).  The mandibles are used to chew the food, while the maxillae are used to help tear food into smaller chunks.  The maxillipeds are used to hold, touch, and even taste the food.

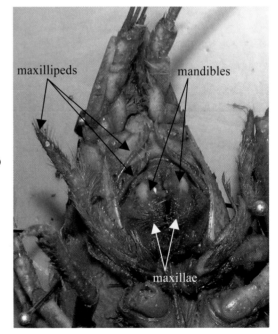

maxillipeds    mandibles

maxillae

# Leg Features

Look at all the legs coming off the crustacean in the drawing on page 123.  The first pair of legs consists of claws called **chelipeds** (kee' luh pedz). "Chel" comes from the Greek word that means "claw," and "ped" refers to feet. Chelipeds, then, are "claw feet." Can you guess what the chelipeds are used for? The crustacean uses its chelipeds to capture prey and to fight.

Most crustaceans have the ability to drop-off their chelipeds. Why would a crustacean do this? Well, if a predator grabbed it by its claw, the crustacean could release the claw and escape unharmed. Unharmed, that is, except for its missing claw. But what's a missing claw compared to a missing life? Besides, a new claw immediately begins regenerating. At the beach, you may run across a crab with a rather small claw. It is probably growing a new one to replace the last one it gave up.

This crab is a decapod, because it has ten true legs.

The next legs are the walking legs, and as their name implies, these are the legs that a crustacean uses to walk. Count the walking legs on the crab pictured to the left. It has four pairs of walking legs. Many crustaceans have four pairs of walking legs and one pair of chelipeds. This makes ten true legs all together. These crustaceans are called **decapods**, because "deca" means "ten." Decapods are the most familiar crustaceans in the ocean. Most shrimps, crabs, and lobsters are decapods.

# Hind Features

Behind the cephalothorax, we find a crustacean's abdomen. Under the abdomen of some crustaceans (like shrimps and lobsters), you'll find little things that look like legs, but they are not true legs. They are called **swimmerets** (swim' uh retz'), and they act as little paddles that help propel a crustacean through the water. In addition, some crustaceans actually have a separate set of gills in their abdomen, and the swimmerets help to push water over those gills. Finally, they are also used by the females to hold eggs. Female crustaceans lay eggs, and instead of just dropping them in the water, they secure them to their swimmerets where the eggs will stay until they hatch.

Every newborn crustacean goes through a larval stage where it looks nothing like the adult. When the crustacean is in the larval stage, it is considered zooplankton and can be eaten by many plankton-eating animals. If it survives the larval stage, it will molt several times, becoming closer in appearance to the adult with each molt. Once it is an adult, it still molts to grow larger and larger until it is fully grown.

On many crustaceans, the abdomen ends in a **tail fan**, which the animal uses for steering while it swims. Crustaceans with a tail fan have an interesting ability: they can swim backwards at very high speed by flicking the tail fan. This is sometimes called "lobstering," and it is typically used when the crustacean feels threatened and wants to make a hasty retreat.

*Tell someone all that you have learned about crustaceans and their anatomy.*

# Lobsters

Lobsters make up some of the biggest crustaceans in the sea. The largest lobster on record weighed 44 pounds. The distance from its tail fan to the tip of its claw was 3 feet, 6 inches. At the same time, some lobsters are so small that they can fit on the tip of your finger.

Though they can be quite aggressive with one another, lobsters are shy creatures and often gather in caves, under rocks, or in crevices during the day. They are nocturnal

Like most lobsters, this Spanish lobster is active at night.

creatures, which means they become active at night. That's when they leave their shelters to look for food.

Lobsters are omnivores, which means they eat plants as well as other animals. They will even consume dead animals they discover, so they are considered scavengers. Lobsters can also be **cannibals** (kan' uh bulz). Do you know what that means? It means they will even eat other lobsters. Interestingly, unlike other arthropods, lobsters (and most crustaceans) have teeth. These teeth aren't in the lobster's mouth, however. They're in the stomach! That's where all their food is chewed.

Lobsters make meals of many creatures, but many creatures also make meals out of lobsters, including octopuses, **nautiluses** (naw' tuh lus ez), and other lobsters. People also enjoy the taste of lobster, which is why you can find tanks of them in restaurants and some grocery stores. Strangely, the lobsters kept in these tanks are not fed. This is because lobsters go into a kind of hibernation and quit eating in extremely cool water. In fact, they can go a whole year without food if the water temperatures are kept really low. Restaurants and grocery stores, then, just keep the water temperature low, and that way, they don't have to feed their lobsters.

If you have ever watched lobsters in a restaurant or grocery store tank, you have probably noticed that their claws are held shut with a band. This is done because, if they were able to use their claws, they would probably tear each other apart, removing claws and possibly other body parts. This is because lobster claws are very powerful, and when forced to be around other lobsters in a small area, a lobster can become very aggressive.

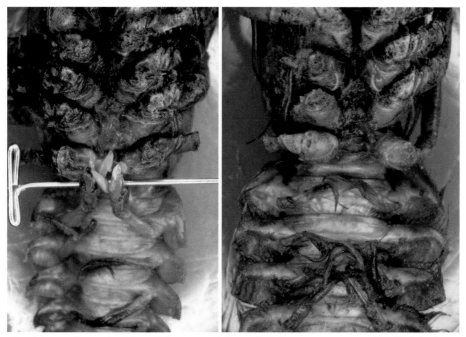

Next time you go to the grocery store or restaurant, see if you can tell the difference between a male lobster and a female lobster. How would you do this? Well, a female lobster has a wider abdomen because it stores all its eggs there. Also, if you can look at the underside of the abdomen, a female's first swimmerets are small and delicate, while a male's are thicker and stronger.

Looking at the undersides of these two lobsters' abdomens, can you tell which is the male and which is the female? The answer is in the "Answers to the Narrative Questions" section in the back of the book.

A female lobster may have as many as 100 thousand eggs under her abdomen at one time! She carries them around for nearly a year before they hatch into larvae and float away as plankton. Of the 100 thousand hatchlings, only about four survive to adulthood. Most will be eaten by plankton eaters, like whale sharks, manta rays, baleen whales, and many other creatures that survive on plankton.

When the larval stage is over, the lobster sinks to the seafloor, where it will spend most of its time hiding from predators. By the time it is five years old, it will have molted about thirty times.

# Crayfish

The crayfish is another kind of crustacean. It is nearly identical to a lobster except for two things. First, a crayfish lives in freshwater (like streams, rivers, and ponds), while a lobster lives in saltwater (the oceans). Some crayfish even live in wet mud, where they burrow down deep to find water, which they need to breathe. Second, most crayfish are smaller than most lobsters. Crayfish are also known as crawdads, crawfish, and freshwater lobsters.

Like lobsters, crayfish are nocturnal scavengers, eating dead, decaying animals, fish, shrimp, worms, insects, snails, and even vegetation. They are truly omnivorous. During the day, they hide under rocks, logs, or in small burrows. Though their color really depends on what they eat, crayfish are usually orange, brown, green, or gray. Adults can also be deep red.

# Crabs

If you have ever found a crab on the beach and decided to capture it, you may know the pain of a crab's powerfully strong, jagged chelipeds. These claws are so sharp that they are able to tear food into tiny bits. Like lobsters, crabs are omnivores.

Crabs are decapods. Do you remember what "decapod" means? It means they have ten legs. Crabs are different from other decapods, however, because of their body design. God created these creatures with pancake-flat bodies that have no visible abdomen. A crab's abdomen is small and flat and doesn't look like a tail. Instead, it is "rolled up" under its body and fused to its cephalothorax. You can actually tell the difference between a female crab and a male crab by looking at the shape of the abdomen. Like lobsters, females have thicker abdomens than males have.

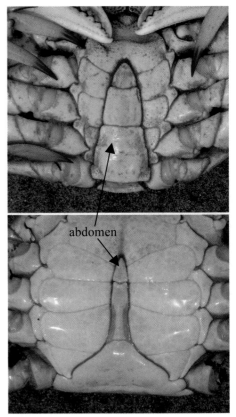

abdomen

Which is the male and which is the female? The answer is in the "Answers to the Narrative Questions" section in the back of the book.

Unlike other decapods, a crab doesn't have swimmerets. Because its body is compact and flat, it doesn't need them to move through the water. It only needs a set of oars. So, God designed a crab's back pair of feet to be paddle-shaped. Some crabs, like the spider crab, don't have any paddle-shaped feet. These crabs never swim; they spend all their time walking along the ocean floor.

If you have ever seen a crab walk, you know crabs don't move along forward as most animals in God's creation do. Their legs don't bend from front to back, but instead bend from side to side. As a result, crabs walk sideways.

When a female crab lays eggs, she holds them to her abdomen in a little mass that looks like a sponge (it's actually called an **egg sponge**) until they are ready to hatch. There are several thousand eggs in one sponge. When they hatch, the babies are very tiny. Many are eaten by fish and other sea animals. The baby crabs will molt five times in one year. On the fifth molt, they are considered adults, even though they are still small. They don't live as long as lobsters do. Most crabs live to be only a year and a half old.

There are really two kinds of true crabs: **marine crabs** and **land crabs**. Land crabs breathe air, while marine crabs breathe underwater with gills. Land crabs don't have a back set of paddle feet, because they don't swim. These two kinds of crabs come in all shapes and sizes. They can be anywhere from a few centimeters to 12 feet across. God made so many different kinds of crabs that it would be impossible to discuss them all here, so I will spend time on the crabs you are most likely to see: fiddler crabs and hermit crabs. I'll also tell you the story of the Christmas Island red crabs.

# Fiddler Crabs

This male fiddler crab thinks the photographer is a threat and is preparing to defend himself and his home (the hole behind it).

A **fiddler crab** is a cute little land crab that lives in mud or sand flats near the beach. You may be familiar with the little mud or sand tunnels made by fiddler crabs along the shores of beaches and marshes. When the tide rises, the crab runs into its hole, quickly building a mud door over the entrance to keep the water out.

The male has one giant cheliped and one small cheliped. Sometimes it's the right cheliped that's the giant one and sometimes it's the left. This giant cheliped is so large that it is sometimes bigger than the whole crab! The reason this crab is called a fiddler crab is because it holds this large claw in front of its body and moves it back and forth in the same way one might play a violin or a fiddle.

Why do you think the fiddler crab needs such a big claw? Well, before I answer this question, I will tell you a little bit about how the fiddler crab eats. First, it opens its mouth, and then it uses its maxillipeds to scoop in a small amount of mud or sand, depending on where it lives. Then, it uses its mouthparts to scrape any food matter (algae, decaying animals or plants, etc.) it can from the dirt it has just stuck in its mouth. After this, it forms the mud or sand into a little pellet, which is called a **feeding pellet**. It then drops the feeding pellet and starts all over again. If you find a fiddler crab colony at the beach, you'll see these little pellets all around the beach.

Okay, since the fiddler crab eats material it can find in mud or sand, why does it need such a big claw? Well, it appears God designed such a big claw on the male fiddler only to attract females! Every two weeks during the summer, fiddler crabs build little round burrows, which are deeper and better than the ones they normally build. The crabs guard these burrows with all their might. In a fiddler crab locale, there are usually hundreds of these little burrows, all lined up only a few inches apart. Each male fiddler crab stands out in front of his burrow waiting for the females to return from eating. The females walk around looking at the different burrows. As they walk past, the males wave their giant chelipeds back and forth. If a female likes a certain male or burrow, she will stop and stare. Then the male begins waving his cheliped wildly. Sometimes the female will then run away. If she stays, however, the male will run toward her, then back to the burrow, and then back to her, over and over again. If she hasn't run away by this time, she will come closer to the burrow. The male will then enter the burrow a bit and start drumming against the side of the tunnel with his cheliped. If the female likes what she hears, she will also enter the burrow, and the two will mate.

# Hermit Crabs

If you've ever had the awe-inspiring experience of playing in a tide pool, you may have seen a shell racing quickly around. If you were brave enough to pick it up, you probably noticed little claws hanging out the opening. Inside was none other than a little hermit crab.

Hermit crabs are unlike any other crustacean. Whereas most crustaceans are covered from head to tail by a hard exoskeleton, the hermit crab is missing part of its exoskeleton! The back part, where its abdomen is located, is soft and squishy. Thus, the minute a hermit crab molts into an adult, it sets out to find a shell in which to live. That way, it can protect its exposed backside.

The rear legs that anchor the hermit crab to the shell are visible as this hermit crab hangs out of its shell.

A hermit crab doesn't make the shell in which it lives. The shell is made by a snail that has either died of old age or has been eaten by another sea creature. When a hermit crab sees an abandoned shell, it tries the shell on for size. The crab moves in if it's a good fit. Later, when it gets too big for that shell, the crab will go looking for a larger abandoned shell.

To get into the shell, the hermit crab scoots in backward, securing itself with its four back legs. The back legs are equipped with hooks that anchor the hermit crab in the shell without any effort. The next four legs are used for walking, while the front two are its chelipeds. As with the male fiddler crab, the chelipeds are not equal in size; one is big and the other is small. The larger cheliped is used for grabbing prey and guarding the entrance to its shell. Like all decapods, hermit crabs have a total of ten legs, including the chelipeds and the rear legs used to anchor itself to the shell.

Hermit crabs are omnivores (eating plants and animals) and scavengers (eating dead animals that they find).  Some hermit crabs are marine crabs (they breathe underwater with gills) and some are land crabs (they can breathe air).  There are about five hundred different species of hermit crabs, and they vary in color, often with patterns like stripes and dots on their bodies.

Today, many people purchase land hermit crabs as pets.  Hermit crab owners often have to buy new shells for their crabs as they outgrow the shells in which they came.

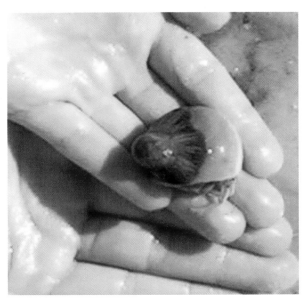

Found in a Florida tide pool, this is a sea anemone (the reddish blob) on the shell of a hermit crab.

Some species of hermit crabs have an amazing symbiotic (sim by ah' tik) relationship with sea anemones.  **Symbiosis** (sim by oh' sis) is when two or more different creatures live together in a close relationship.  These hermit crabs carry sea anemones around with them, stuck to their shells.  In fact, they will often pick up sea anemones and plant them right on top of their shells, and when they move to another shell, they'll take their sea anemone with them!

This symbiotic relationship benefits both the hermit crab and the sea anemone.  The sea anemone eats food particles left by the crab, and the crab is protected from its enemies, particularly octopuses, because the sea anemone has potent stingers that can paralyze animals that try to attack it or its crab-host.

# Creation Confirmation

Symbiosis is a broad term that covers several different types of close relationships between God's creatures.  When the creatures in a symbiotic relationship each help the other, the symbiosis is often called **mutualism**, because the creatures mutually benefit from the relationship.  This is the case with the hermit crab and sea anemone.  The hermit crab benefits by being protected from predators, and the sea anemone benefits by eating food that the hermit crab finds.

Mutualism is another amazing testimony to God's wonderful design.  How does the hermit crab know to pick up the sea anemone and put it on its shell?  How does it know to transfer the sea anemone to its next shell?  How does the sea anemone know to stay on the shell?  Why doesn't it pull itself off the shell and go back to the familiar stationary perch?  The hermit crab and sea anemone know what to do because they have been *made for each other*.  God arranged it all in advance so that the relationship works.  As you study God's creation, you will find many, many examples of mutualism.  In fact, I will tell you about another example later on in this very lesson.  Each time you learn about mutualism, remember that it works because the creatures involved were made for each other.

# Crabs for Christmas?

Before moving on to learn about shrimp, there is one more crab I must mention: the red crab of Christmas Island. This island is in the Indian Ocean near Australia, and it holds a remarkable land crab that does a remarkable thing. You see, at certain times during the rainy season (usually sometime in October, November, or December), more than *100 million* red crabs that have been living in the forest come out of hiding and head right toward the ocean. The males come out first, and the females follow a few days later. But first, all of them must pass through the city. Every year, without fail, the crabs come and take over the city for a week or two. The city is simply overrun by them. Millions of them march down the streets and sidewalks. They enter buildings, homes, and businesses; they cover driveways, lawns, and parks. From a distance, some of the streets on Christmas Island look like they are covered in a giant, moving red carpet.

This is a common sight on Christmas Island during the wet season. Because of this, signs are often posted asking drivers to avoid using the roads, if at all possible.

The people who live on Christmas Island consider the crabs a bit of a nuisance. They sweep them out of their homes and even drive right over them in the streets. The government tries to close some of the streets to protect the crabs, and the people are asked to avoid driving on other streets if at all possible. Nevertheless, about a million of the 100 million crabs are killed on the way down to the ocean. Their chelipeds are so sharp that a car driving over a crab can actually get a flat tire!

You may be wondering why the red crabs do this every year. Well, a red crab's eggs must hatch in the ocean, because the crab larvae breathe with gills. Thus, the parents go down to the beach to mate so that the eggs can develop in the female's egg sponge. When the eggs are ready to hatch, the mother releases them into the water so that the larvae can breathe in the water with their gills. As the larvae mature, they eventually develop lungs and come out of the ocean, heading into the forests on the island. In the forests, they eat mostly leaves and fruits that have fallen on the ground.

Why do they choose a time near Christmas to do this? Well, it turns out that the difference between high and low tide at Christmas Island is lowest during this time of the year. If there were a large difference between high and low tide, the female could get trapped underwater if she were releasing her eggs as the tide advanced. That would kill her, since she cannot breathe underwater. For the thousands of years that the crabs have been making this journey, God knew that the tide was safest around Christmas, so He compelled His creatures to go to the beach at the right time. Amazing!

# Shrimp

A shrimp looks a lot like a lobster.

Shrimp, for all practical purposes, look like miniature lobsters. Even the name "shrimp" has come to mean "small." They range in size from several inches long to almost too small to see. They come in many colors; some are even transparent.

Although most shrimp eat plankton and small dead creatures that are floating in the water, **mantis shrimp** eat worms, fish, and other shrimp. Typically, a mantis shrimp will burrow down into the sand or mud and shoot out its cheliped to spear animals that swim past. Another kind of shrimp, the **pistol shrimp**, catches prey by snapping its cheliped in such a way as to make a really loud popping sound. This stuns the prey, making it an easy catch for the pistol shrimp.

# Symbiotic Shrimp

Shrimp are some of the most symbiotic animals in the sea. They actually clean the teeth of many fish in the sea! This mutualistic relationship helps both the fish and the shrimp. The shrimp eat the stuff on the fishes' teeth, giving them food, and the fish get clean teeth in the process! This is important, because if teeth aren't cleaned, they will rot and fall out. You brush your teeth to make sure this doesn't happen. Many fish use cleaner shrimp as their toothbrush!

The red and white shrimp in this picture is cleaning the eel's teeth.

In fact, many shrimp set up cleaning stations. Fish visit the station, and they line up, each waiting their turn to have their teeth cleaned. When a fish gets its turn, the fish opens its mouth wide. A cleaner shrimp scuttles about inside the mouth, eating whatever it finds lodged in the fish's teeth. The shrimp is not afraid that the fish will chomp down and eat it, because the shrimp knows that it has been made for the fish. In the same way, the fish knows that the shrimp is helping it, so it will not hurt the shrimp. What a beautiful picture of God's creation — animals designed to help other animals!

# Shrimp-like Crustaceans

A single krill (top) is very small, but when many group together, they can turn the water red or pink (bottom).

**Krill** are like tiny shrimp, only as big as 2 inches when fully grown. As you can see from the picture on the top left, they look a lot like shrimp. They are mostly transparent, with light pink or red markings on their exoskeletons. Can you see an immediate difference between shrimp and krill? Count the walking legs on the krill. As you can see, krill are not decapods, so they aren't even really shrimp. They are crustaceans, however. They just belong in a different order from shrimp.

Most krill live in the cold Antarctic waters, though a few can be found in the open ocean. A lot of sea creatures are dependent on krill for survival. In fact, if the giant blue whale had no more krill to eat, it would probably become extinct. Whales are not the only animals that love to feed on krill. Birds, fish, and seals also enjoy these tasty ocean crustaceans. Because so many other animals depend on them for survival, krill are considered a **keystone species**. As you learned in your first-year zoology course, if a keystone species were to become extinct, it would cause a lot of animals to die out as well.

Some krill spend their days in the dark depths of the ocean, safe from birds, seals, and whales. They then swim to the surface each night to eat. Other krill live in shallow waters near the shore. Because they are small, the only food they can eat is other plankton. The krill that live in the coldest regions prefer phytoplankton, which are typically green. When you look at these krill, they look like they have bright green digestive organs, but this is really just the color of the phytoplankton they ate. Krill can actually fast (go without eating) for up to two hundred days. So, if there is no food about, they can usually survive until food is available. As they fast, they actually shrink in size!

Krill reproduce in huge numbers. A female may have more than five thousand offspring each year. Unlike shrimp, which are generally loners, krill prefer to swim with other krill in huge swarms. This could be for safety's sake, or it could be that God has them swim together so that a giant baleen whale can get a lot to eat in a small area. As the picture on the previous page shows, krill gather in such huge numbers that when a swarm of krill is nearby, it looks like an orange cloud is blowing through the sea. These huge swarms amaze people and delight the animals that eat them.

*Explain in your own words what you learned about krill.*

# Barnacles

Have you ever seen **barnacles** like the ones in this photo? You usually find them attached to something, like a whale, the side of a pier, or rocks on the shore. Perhaps you thought you saw just a bump of dirt or sand collected there, but that is not so. Inside that little "bump" is a crustacean — a little crab-like creature. The part of a barnacle that you usually see is its exoskeleton. Many people call it a shell, but it is not. It is an exoskeleton, much like that of a crab.

When a barnacle is underwater, it sticks its twelve feathery legs out of its exoskeleton, grabbing any plankton or debris

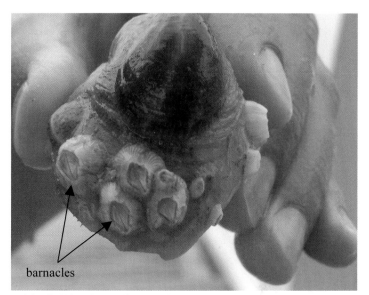

barnacles

This shell has barnacles attached to it. The "door" to each barnacle's exoskeleton is shut, so you can't see what's inside.

that happens to float past. It brings the food particles back inside the exoskeleton, placing them in its tiny mouth. Do you remember what creatures that eat like that are called? They are called filter feeders. When a barnacle is out of the water, such as when the tide goes out, it is able to fold up into its exoskeleton and shut the "door" so it makes a watertight seal. It will not open the door again until it is once again covered with water.

Barnacles will attach to anything in the ocean. They make a kind of glue that's so powerful that it is very hard to pull them off a surface once they stick to it. It's God's Super Glue! Because of this, barnacles are a nuisance to boats. When they stick to a boat, they increase the water drag the boat experiences, which makes the boat waste fuel. Some boats are painted with a barnacle-repelling paint to prevent this from happening. Boats that don't have such paint must have the barnacles scraped off from time to time. Next time you are at the beach, try to find barnacles on rocks, piers, or shells.

# Horseshoe Crabs

The first time you see a **horseshoe crab**, you might be frightened. This creature looks dangerous with its giant, armored shell and long, spiky tail, but it's actually harmless. The horseshoe crab, sometimes called the king crab, is misnamed because it's not a crab at all. In fact, it's not even a crustacean, because it does not belong to class Crustacea. It is, however, an arthropod, because it is classified in phylum Arthropoda. Thus, it has some things in common with crabs, because crabs are also in phylum Arthropoda. Horseshoe crabs, however, are in class **Xiphosura** (zye' fuh sir' uh).

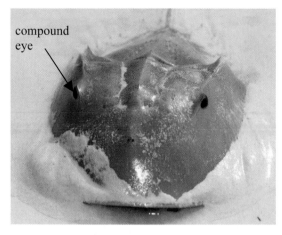

compound eye

Although it might look dangerous, a horseshoe crab is harmless.

On its dome-shaped carapace, the horseshoe crab actually has four eyes. It has two compound eyes that protrude from the sides of the carapace, and two simple eyes that are on the front of the carapace. Its eyes are positioned high on the carapace so that it can see when it buries itself in the mud, which is the thing it enjoys doing most.

A horseshoe crab has twelve legs: five pairs of walking legs and a set of tiny chelipeds. The long tail, though it looks frightening, isn't used as a weapon at all. It would never hurt you with it; instead, it's used for steering and for flipping the body over if it gets turned upside down. So, if you see a horseshoe crab, you can actually pick it up and study it a bit before returning it to the ocean.

Horseshoe crabs can be found all over the eastern coast of the United States, preferring beach habitats. Buried under the mud or sand, they like to eat small clams, crustaceans, and worms. All over the coast, you will find their shells washed up on shore. Some of those shells are rather large, as the horseshoe crab can grow up to 2 feet long and a foot wide!

Horseshoe crabs will go through a lot of molts to reach that grand size.

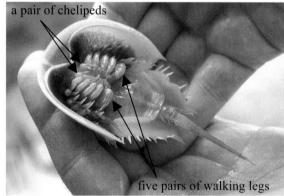

a pair of chelipeds

five pairs of walking legs

Like all horseshoe crabs, this small one has twelve legs.

Unlike female crustaceans, the female horseshoe crab does not keep her eggs on her body. Instead, she goes to the beach and digs holes in the sand. She buries her eggs in the hole and then covers them with sand. When the larvae hatch, thousands are gobbled up by the birds and other predators. However, don't fear. Even though many horseshoe crab larvae are eaten, many will also escape the feeding frenzy and make it out to the ocean to become adults.

# Trilobites

What we know about **trilobites** (try' luh bytz) comes from their fossils, because as far as we can tell, they are extinct. At one time, however, the ocean floor was crowded with these little creatures that could roll up in a little ball like a pill bug. There were many different species of trilobites. Some were tiny, not even ¼ of an inch long. Others were over 2 feet long. Most trilobites, however, were small, between 1 and 3 inches long. Would you believe that each year, scientists find the fossils of at least one new species of trilobite?

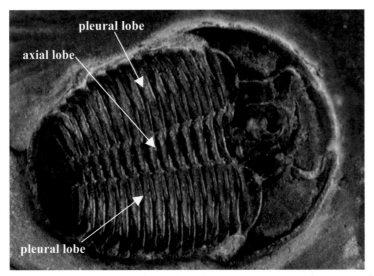

pleural lobe

axial lobe

pleural lobe

It is easy to see the three lobes on this trilobite fossil.

A trilobite had a hard exoskeleton covering the top part of its body, but its underside was soft and fleshy. Its whole body was divided into three lengthwise sections, or **lobes**. These lobes are where the trilobite gets its name. Do you know what the prefix "tri" means? If you said three, you are correct. The trilobite had a center lobe called an **axial lobe** and two lobes to either side of the center called **pleural** (ploo' ruhl) **lobes**. That makes three lobes, which is why the animal is a tri-lobed-ite, or trilobite.

Most trilobites had two antennae and two very large, compound eyes. Some trilobites had no eyes, and a few had eyes that towered over their heads on long stalks, like the eyes of snails. Trilobite eyes were unlike the eyes of any other creature because of their special, complex design. They were multifaceted, having many lenses, similar to the eyes of insects. That means that the trilobite had the ability to see in many directions at the same time. But a trilobite's eyes were not made of living cells, like our eyes are. Instead, they were made of a see-through, crystal-like substance. Because trilobite eyes were basically tiny "rock crystals" and not soft flesh that would decay after they died, some trilobite eyes have been perfectly preserved as fossils.

## Try This!

Hold your finger in front of your nose and look at it. Now without moving your finger, look across the room at something else, like a picture or a lamp. What you just did is called focusing. Try to focus on your finger and the object across the room at the very same time. Can you do it? No, you cannot! If your finger is in focus, the lamp will be fuzzy. If the lamp is in focus, your finger will appear fuzzy. You see, eyes made of soft tissue, like yours, can see things that are nearby and also see things that are far away because your muscles change the shape of the lens slightly. But they cannot focus on two things at the same time. However, the crystal eyes of a trilobite could see things far away and things nearby equally well at the very same time! Exactly how the trilobite did this is very complicated, but it centers on the fact that each trilobite lens was actually made of two different types of material. The trilobite's crystal eyes are amazing feats of engineering. In fact, in his book called *Trilobites*, Dr. Riccardo Levi-Setti (a world-renowned expert on the subject) said that the design of a trilobite's eyes could qualify for a patent. Of course, the Designer who made them does not need a patent, as the entire world is His!

The many legs of the trilobite were long, slender, and jointed, helping it plow its way through the sediment on the seafloor and hold itself steady in the ocean currents. The legs worked together in rhythmic waves along its body, just like a centipede's legs do. Do you think your mother would allow you to eat with your feet?  Well, the trilobite's legs moved in such a way that they channeled tiny bits of food into grooves that ran the length of its body and led to the mouth on the underside of the head. The movement of the legs kept the food constantly moving in the direction of the mouth.  If you choose to raise triops (try' ops) at the end of this lesson, you will notice that they have legs that do the very same thing.

So what caused these little creatures to become extinct, and why did so many trilobites become fossilized at the same time?  No one knows for sure, but it would seem that a great change came to their world, which was the bottom of the sea.  Perhaps the trilobites were simply wiped out by the cataclysm of the worldwide Flood.  The Bible says that God opened the "fountains of the deep" in order to raise the floodwaters.  That would surely change the ocean floor drastically.  Perhaps those changes were just too much for trilobites.  Whatever caused their extinction, it had to be something that occurred worldwide, because trilobite fossils can be found almost everywhere.

Since trilobite fossils are fairly easy to find, they are not expensive, like many other fossils.  If you live in an area where trilobites are found, you might want to begin a collection of your own.  Visit a rock shop or join a collecting club to find some nice specimens.  Use a magnifier to observe them closely.  If you are lucky, you might even get the chance to see those beautiful crystal eyes for yourself!

# What Do You Remember?

What does the word "arthropod" mean?  What is an exoskeleton?  How does a crustacean molt? How do antennae help crustaceans?  What are maxillipeds?  What are chelipeds?  What are some of the uses of swimmerets?  How long can a lobster live?  How are crabs different from lobsters?  What is the symbiotic relationship between shrimp and fish?  What does it mean to be a keystone species? Which crustacean is a keystone species?  Where do barnacles live?  Where do horseshoe crabs lay their eggs?  What kind of eyes did trilobites have?

# Notebook Activities

For your notebook, draw a picture of a lobster. Label the body parts, as done on page 123. Also, draw a picture of a crab and list the differences between crabs and the "typical" crustacean you drew before.  In addition, make up a conversation that might take place between a mother crab and a mother horseshoe crab. Have them discuss the differences in how they lay and care for eggs. Perhaps the mother crab is tired from carrying the eggs around all the time. Perhaps the mother horseshoe crab misses her eggs and wishes she didn't have to leave them. Be creative! There are pages for the above activities in the *Zoology 2 Notebooking Journal*.

# Project
## Make an Animal Quiz Game

**You will need:**

♦ Markers

♦ A blank file folder

♦ A set of index cards

♦ Dice

♦ Animal game pieces for each player (You can make these or use small plastic animals.)

Open the file folder and use markers to draw a game board with squares (or any other shape) in different colors that begin at a place called "start" and end at a place called "finish" or "end." You can make your track any way you wish; you can draw animals or simple shapes. You can make your path curvy or make it go around the board in a square, like monopoly. Use your imagination to make your own design.

After that, begin making your Animal Quiz cards. On one side of each index card, write a question with multiple choice or true/false answers. The players will have to answer these questions correctly in order to move forward. Write the correct answer in very small letters at the bottom, beneath the question. The person being asked the question will not be able to see the card when it is his or her turn, so another player will have to read the card to him or her.

Use the information you have learned in this book and other places to come up with the questions. Although we only studied crustaceans in this lesson, you can make your animal cards about anything you have learned so far. You should create a lot of these cards so that each time you play, you will have different questions to answer. Have every member of the family make up cards so that one person doesn't know all the answers. As you learn more, you can continue to add to your quiz card collection. Examples of questions you might have are:

*Decapod is the word for:*
   *a. A very old crustacean*
   *b. The burrow home of a fiddler crab*
   *c. The top deck of a shrimp boat*
   *d. Crustaceans with ten legs*
(The answer is d)

*Answer true or false:*
   *Although krill are very small, they tend to eat animals that are much larger than they are.*
(False)

To play the game, turn all of the quiz cards upside down. The youngest player (Player 1) will begin first. She will roll the dice. Then the player next to her (Player 2) will pick up an Animal Quiz Card and ask the question on the card without showing the answer. If Player 1 answers correctly, she can move the number of spaces indicated on the dice she rolled. The player who reaches the finish first wins!

# Ocean Box

Create some crustaceans for your ocean box. You should have at least a crab and a lobster or shrimp. We used chenille sticks (pipe cleaners) to make the legs and antennae for the lobster. Since these creatures live at the bottom of the ocean, they should be at the bottom of your ocean box.

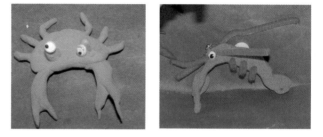

# Experiment

I mentioned two shrimp-like creatures that are fun to buy and grow in your own home: Sea-Monkeys and **triops**. These two creatures are tiny crustaceans that can stay inside their eggs for many years in a totally dry state. In fact, triops eggs won't hatch unless they have been dried. They thrive in areas that dry up and then flood throughout the year, like rice fields and shallow swampy areas. Sea-Monkeys are really **brine shrimp** (not true shrimp, but a shrimp-like animal) and prefer saltwater. Triops, sometimes called **tadpole shrimp** (also not true shrimp), look similar to miniature horseshoe crabs and prefer freshwater. Both are omnivorous.

You can purchase Sea-Monkeys or triops just about anywhere, especially places that sell educational science products. If you visit the course website that was given in the introduction to this book, you can find links to Internet sites where you can buy them. Simply follow the directions on the package for hatching your crustaceans. They usually only live a few months, so enjoy them while they last!

See if you can train your crustaceans to swim to the top when you tap on the tank. Do this by always tapping on the tank just before you feed them.

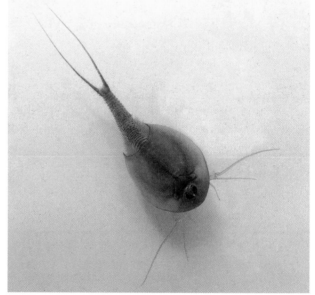

This triops looks a bit like a horseshoe crab, and it behaves a bit like a trilobite.

# Lesson 9
# Mollusks

I was once in a large grocery store where clerks were offering samples of different things to eat. One man had a bowl full of shells that were half opened; inside was something that looked like rubber. When I ate this little piece of rubber, it was very chewy and tasted like fish and spices. I had just eaten a **mollusk** (mah' lusk), more specifically a **mussel** (mus' uhl).

Many people enjoy the taste of mussels.

Mollusks, which belong to phylum **Mollusca** (mah' lus kuh), are soft, slimy creatures, with squishy bodies. The snails and slugs in your yard are mollusks, as are the animals in the sea called sea snails, oysters, clams, mussels, sea slugs, and many others.

Interestingly enough, the crustaceans you learned about in the previous lesson and the mollusks you will learn about in this lesson are often called "**shellfish**." This is odd because neither crustaceans nor mollusks are fish, and as you should recall, crustaceans don't really have shells. They have exoskeletons. Nevertheless, if you see shellfish on the menu in a restaurant, you know it means either crustaceans or mollusks. In any case, people all over the world love to eat these animals.

The shell on the left was made by a sea snail, which you can see when you turn the shell over (right).

If you have ever found a shell on the beach or bought one in a store, you can thank a mollusk for it. You see, every seashell you find on the beach was made by a mollusk. Mollusks even make the shells in which hermit crabs live. So, every shell you find was once the home of a mollusk that made it. However, not all mollusks make shells. For example, the octopus is a mollusk that doesn't make a shell.

Though you may not think much of them, sea snails are actually very interesting creatures. A sea snail is similar to the snails in your yard, except it has a totally different life. It leads a more dangerous life, fraught with the risk of being eaten by other mollusks! In fact, the main predators of sea snails are other sea snails.

Are you wondering how a mollusk can make such a complicated and beautiful shell? Well, inside the body is an organ called the **mantle**. God equipped this mantle with special chemical properties that convert calcium and other minerals into a shell. Every shell, from a simple clamshell to the elaborate conch (pronounced kongk) shell, is made by the mantle of a mollusk. In this lesson, we will study the groups of mollusks called **gastropods** (gas' truh podz') and **bivalves** (by' valvz). Slugs and snails are examples of gastropods, while clams are bivalves. In the next lesson, we'll study the group of mollusks called **cephalopods** (sef' uh luh podz'), which includes octopuses and squids.

# Bivalves

Bivalves like clams and oysters have shells made of two halves which open up. On the beach, you often find only one of those halves.

When you are hunting for seashells on the beach, you will most likely find shells belonging to **bivalves** — or rather, half of a shell belonging to a bivalve. Bivalve shells have two shell halves, which is exactly what "bivalve" means. These shell halves are concave and shaped like a fan or an oval. Piles of them will wash up onto the beach after a storm. If you dig deep into the sand, you can actually pull up living bivalves. They are fun to observe if you place them in a sand pail or bucket of water.

Bivalves include mussels, scallops, cockles, oysters, clams, and bizarre little mollusks known as shipworms. The two shell pieces of a bivalve are hinged together at the tip. A bivalve can use its muscles to open its shell or keep its shell tightly closed.

When a bivalve is underwater, it keeps its shell partly open so that it can extend out tubes that are called **siphon** (sye' fun) **tubes**. It uses these tubes so that it can eat, breathe, and excrete wastes. One siphon tube, called the **incurrent** (in kur' unt) **siphon,** takes in water. This water flows over the bivalve's gills so that it can pull out the oxygen that has been dissolved in the water. This allows the bivalve to breathe. Also, food particles that are floating in the water (such as plankton) are removed so that the clam can eat. The other tube, the **excurrent** (ex kur' unt) **siphon**, expels wastes.

excurrent siphon

incurrent siphon

This clam is using its siphons to breathe, eat, and get rid of wastes.

How many whelk holes can you find in the shells pictured above?

Many animals feast on bivalves. For some, it's the meal they prefer over all others. A walrus can eat thousands of clams in one eating period. Otters also seek out clams each day to satisfy their appetites. Starfish love to feast on clams as well. Even so, the most common predator of bivalves is the **whelk**, which is a sea snail. The bivalve's ability to clamp shut with amazingly powerful muscles does not deter the whelk one bit. You see, a whelk is equipped with a little razor that can drill a hole right into the clamshell. Once the hole is drilled, the whelk sucks out the bivalve. This happens so often that you are likely find clam shells with holes in them on the beach. Blame those holes on whelks!

## *Bon Appétit,* Bivalve

I already told you *how* a bivalve eats, but *what* does it eat? A bivalve inhales water that has plankton and debris floating in it. Any particles that are small enough to fit inside the hole of the incurrent siphon will enter the bivalve. When the floating material comes in, it gets stuck in mucus that coats the surface of the bivalve's gills. The food is moved down to the mouth, which is on the other side of the siphon. Food is digested in the bivalve's stomach and intestine, and everything that isn't digested exits through the excurrent siphon. Bivalves take in many bacteria and other unhealthy or unclean products, removing them from the water. Thus, bivalves are part of an elaborate system created by God to clean the water for us. This means that if you want to eat a bivalve, you need make sure it came from clean water that is free from contaminants.

## Burrowing Bivalves

Can a bivalve move? Yes, it can. In fact, a bivalve actually has a **foot**. It can open its shell and extend its foot, using it to crawl along the seafloor. In addition, many bivalves use their foot to burrow deep in the sand or mud. A few bivalves can actually bore into hard surfaces, like wood or rock. A few bivalves even have the ability to swim. A scallop, for example, can clap its shell open and closed to move short distances in little bursts.

Birds are aware of burrowing bivalves. Sea birds scour the seashore for bivavles. They poke their beaks into the sand, hoping to find a bivalve that will make a tasty meal. Since different bivalves burrow to different depths, some birds have beaks long enough to get to one type of bivalve, while other birds have beaks long enough to get to other types of bivalves.

# Clams

Have you ever been to the beach and noticed little holes that appear in the sand each time the waves retreat? You probably know that an animal makes the holes, but what kind? Most likely, clams made those holes. A **clam** has a burrowing foot like the one in the image on the right. It can use its foot to dig down into the sand or mud to hide. Scientists classify clams by how far down they dig and into what kind of surface they dig. Some burrow less than an inch under the sand, while others dig several inches below the surface. In order to breathe and obtain food, they leave holes in the sand where water can come in, carrying oxygen for the clams to breathe and floating food particles

Although you might think that this clam is sticking its tongue out at you, it is actually sticking its foot out of its shell.

for the clams to eat. When you stroll along the beach, take note of the little holes that appear here and there. Most likely, clams reside down below.

Some clams can actually burrow into coral or rock. They secrete a substance that weakens the material and then begin digging with the edges of their shells to form a hole. Some clams, such as shipworms, use their shells to burrow holes in ships or piers. Shipworms are known as the termites of the sea, because they can cause terrible damage to wooden things like ships and piers. Ships sink and

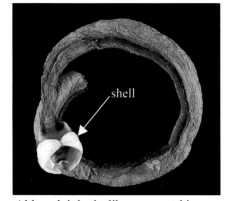

Although it looks like a worm, this shipworm is a clam. Its shell does not contain its body, however.

piers collapse because of all the holes they dig into the wood. In fact, some believe shipworms caused the Spaniard's defeat when the British conquered the Spanish Armada in 1588. Piers have recently collapsed in New England due to an increase in shipworm numbers. Why did the shipworm population suddenly increase? Well, after the government instituted a very successful program to clean up the water, these clams suddenly appeared in great numbers. After the oily, chemical-filled water was cleaned up, the population of microscopic life grew, giving the clams a lot of food. The clams thrived and ruined the piers, which had been fine for hundreds of years in the polluted water!

There are probably more than 15 thousand different species of clams in the world. They range in size from tiny, smooth clams to mid-sized, ridged clams to giant clams. Of course, the giant clams are not really giant. They grow to be about 4 to 4 ½ feet across and weigh about 500 pounds. While that's not "giant" to you and me, it is giant for a clam!

Next time you find a clam shell, look for rings that go all the way around the shell. You can actually tell the age of the clam by counting those rings. Each ring represents a new layer of shell that was added each year as the clam grew. This new layer creates a ring, similar to how a tree forms rings when it grows each year. A clam that has five rings is five years old. You may even notice that some rings have more distance between them than others. Those are years in which the clam grew larger than usual. Perhaps it ate more that

Each of the rings you see on this clam shell was formed in a year. Counting the rings will tell you how old the clam was when it died.

season, or perhaps the climate stayed warmer for a longer time, so there was more for the clam to eat over a longer period of time.

# Clinging Creatures

If they are not forcibly removed, **oysters** spend their entire adult lives in one spot, clinging to rocks, piers, or other structures in the water, even when powerful waves slam against them continuously. **Mussels** also attach themselves to a spot, but they will move from time to time. These bivalves use tough fibers called **byssal** (bih' suhl) **threads** to attach themselves to any suitable surface. These byssal threads act like natural Super Glue. In fact, the creatures are so powerfully attached to the surface they choose that they are nearly impossible to remove, unless you use a tool. To this day, human science cannot produce such a strong glue that is not degraded by saltwater.

Mussels are usually smooth black, bluish, or gray in color, growing from 1 to 9 inches. When a mussel is born, it is in its larval form. These larvae float around in the water until they find a place to settle. While in its larval stage, the mussel is known as a **spat**. The spat will grow and mature until it becomes an adult. During high tide, it moves around by pushing out byssal threads that stick to rocks and other structures. Then, it pulls itself along using these threads. It searches around, scraping microscopic creatures from rocks and other surfaces. When the tide goes back out, it returns to its spot and closes its shell.

Although many people like the taste of mussels, oysters are a much more popular dish. Because oysters are eaten all over the world, many fishermen fish for oysters, pulling them out of muddy flats by the thousands or using pick axes to pop them off the rocks to which they cling. Oysters are harvested all year long. Folklore held that oysters should not be eaten except during months that have an "R" in them; that is September through April. That was once true, because lack of refrigeration during transport in the warmer months caused the oysters to spoil. Thus, only in cooler weather were they unspoiled. Now, however, with proper refrigeration, oysters can be harvested and eaten all year long.

Oysters usually have a wrinkled look to their shells, with one valve of the shell larger than the other. The larger one fits like a lid over the smaller one. An oyster cements its smaller valve to structures like rocks or piers, and the other valve can open to allow the oyster to take in water. Oysters then filter food out of the water that they take in, like clams.

This is a typical oyster, with a wrinkled shell and one valve that is smaller than the other.

One very strange thing about oysters is that most of them are both male and female at the same time. Do you remember hearing about this before? Certain fish are the same way. Can you remember what we call animals like this? We call them hermaphrodites.

# Pearls

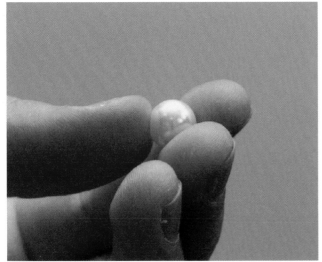

Pearls like this one can be highly valued, but in fact, they are just bits of dirt encased in shell.

Have you ever seen a pearl? Pearls are expensive stones that some people wear in necklaces, rings, and earrings. They are small, shiny balls that can be tiny (the size of a peppercorn) or very large (the size of your fist). Can you believe that oysters, mussels, and clams actually create these pearls?

A pearl forms when something like a grain of sand gets inside the bivalve shell right between the mantle and the shell. Do you remember what the job of the mantle is? It makes the shell. Well, the mantle reacts to this grain of sand by secreting layers of shell material around it. Soon the object is completely encased inside layers of shiny shell. Over the years, these layers of shell continue to form over the object, growing larger and larger. That's all a pearl is! It is a piece of dirt, sand, or something like that, which has been covered by many layers of shell over the years.

The world's largest pearl was found in 1934. It's a 14 pound pearl called the Pearl of Lao-tze, and it is about the size of a basketball!

There are two kinds of pearls: saltwater pearls and freshwater pearls. Saltwater pearls come from oysters that live in the oceans. These pearls are usually perfectly round and white and are the most expensive kind you can buy. Freshwater pearls are found in mussels that live in rivers, lakes, or ponds and tend to be irregular and more varied in color.

The making of a natural pearl by an oyster is accidental, and it's also very rare. So, today, there are places where oysters are grown and tiny beads are purposely inserted into the oysters to begin making pearls. The pearls are then harvested after a few years. These are called **cultured pearls**. Cultured pearls and natural pearls are both made by the mollusks in the same way; however, the cultured pearl is started in an unnatural way.

# Swiftly Swimming Scallops

One special, fan-shaped mollusk is the fastest mover of all the bivalves. It is called the **scallop** (skah' lup) because it has raised ridges on its shell that typically form a wavy, scalloping pattern. It is the fastest moving of the bivalves because it can actually "swim." It swims by opening its shell and then clapping it closed quickly. This pushes the scallop through the water in short bursts of speed.

Notice the eyes that form two rings of dots around this scallop.

Scallops have many tiny eyes that form rings around the animal. They are very sensitive to changes in light levels and the movement of shadows. This allows scallops to see danger approaching so that they can escape.

*Tell someone in your own words what you have learned about bivalves.*

# Sea Snails

While you might easily find a bivalve stuck to the side of some surface, gastropods are harder to find. Gastropods are snails that generally live inside coiled shells that many people enjoy collecting. If you see one on the beach, it will probably not contain a snail. However, you might find a hermit crab residing within.

Snails are called gastropods because "pod" means "foot," and "gastro" means stomach. These animals, then, are "stomach foot" animals. In other words, their whole body is made up of a coiled mass of organs (like the stomach) sitting on top of one large foot. Imagine if your body was nothing more than your internal organs sitting on top of a foot!

This moon snail is an example of a gastropod that lives in the ocean. In other words, it is a sea snail.

Snails have a good sense of smell. Sea snails, for example, can smell food from hundreds of feet away or even farther if they catch a whiff from a sea current. Sea snails also scurry along much more quickly than the slowly-trudging land snails in your yard. When they smell a dead fish in the shallow waters near the shore, they'll even hitch a ride on the waves to get to it quickly. Within minutes, the dead fish is covered with whelks, conchs, or other sea snail scavengers in the area. Many sea snails are, indeed, scavengers that eat dying or dead, decaying creatures. Others are powerful predators that move around the sea, preying on bivalves and other animals.

Like most mollusks, gastropods have an organ called a **radula** (rad' yoo luh). It is a jagged organ that has many denticles. Do you remember what denticles are? Sharks have them on their skin. They are teeth-like spikes. This means that a gastropod's radula is like a jagged knife. Some gastropods use their radula to scrape algae off rocks and into their mouths. Others use it like a drill so that they can make a hole in an animal's shell and eat the animal within. Do you remember we mentioned finding a shell on the beach with a hole in it? The hole tells the tale of the bivalve's sad demise when a carnivorous sea snail seized it for a snack.

Do you remember that starfish like to eat mussels and clams? Well, a certain sea snail, the giant triton, has its revenge on sea stars. It clings to any starfish it can find and cuts into it with its radula, eating the soft insides.

operculum
This gastropod uses its operculum as a door to seal off its shell.

Look at the sea snail pictured on the right. Notice the thick pad called the operculum. It's right on the end of its foot and is used like a door. When the snail retreats into its shell, it can close this door for protection and to hold in moisture. If you ever find a sea snail on the beach or in a pond, turn it over. Is the door open or closed? If the sea snail is still in the shell, it has probably squeezed its body into the shell and shut the operculum door behind it.

*Tell someone everything you remember about gastropods.*

# Conchology

Sea snails can be found in the deepest, darkest parts of the oceans, where no light has ever penetrated. Most, however, can be found right on the shoreline or washed up on the beach. Since each kind of gastropod makes its own kind of shell, there are *a lot* of different kinds of gastropod shells to be found. **Conchology** (konk awl' uh jee) is the study of shells. We get the name from one of the most popular gastropod shells, the conch shell. Let's explore and learn about some of the different gastropods and their shells.

# Conchs

Notice how large the lip of this conch shell is.

**Conchs** produce what is probably the favorite shell of most conchologists, the conch shell. The **queen conch** produces the largest shell, and it is quite beautiful. Many conch shells are large and have beautiful color patterns and designs. The opening of a conch shell has a wide, pearly colored lip that flares outward. Sometimes, this lip is larger than the entire shell. The shells have a very distinct notch at the front end, called the **stromboid** (strom' boyd) **notch**. A conch has eyes on the ends of its antennae, and one of the conch's eyes can protrude right out of that notch.

Conchs are generally found in warm waters, and they generally stay near the beach where they graze on algae and debris. Young conchs can bury themselves in the sand when they are in danger. One of their greatest dangers comes from rays, which can crush a conch shell with their powerful jaws. Conchs are also in danger of being harvested by people who either want their beautiful shells or want to eat them!

For many thousands of years, the conch shell has been used as a trumpet or horn. Typically, the small tip of the shell is removed to form a mouthpiece. When a person blows through this mouthpiece, a beautiful, loud tone that can carry a long way is produced. The pitch can be changed by inserting a hand into the shell and varying the position of the fingers.

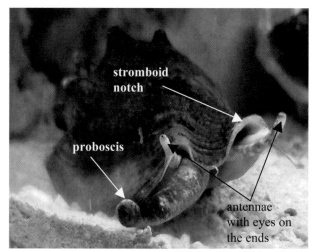
This conch has both eyes and its proboscis sticking out of its shell.

# Whelks

A whelk shell doesn't have the extended lip that a conch shell has.

Many people mistake **whelks** for conchs, but a whelk shell doesn't have the extended lip that a conch shell has, and it is not nearly as thick as a conch shell. Nevertheless, whelk shells are very beautiful and grow up to 10 inches long, though most are less than 2 inches. Whelk shells are typically spiral shells, some with pointy spikes, knobs, or other interesting features on them.

Many whelks are ferocious predators. These carnivores slide along the seafloor, seeking clams, oysters, worms, and other gastropods to eat. A whelk's radula has a long stalk, called a **proboscis** (pro bah' skus). Some whelks use this proboscis to probe into other gastropod shells or to pry open bivalve shells, using the point of the shell like a crowbar. Others will use it to excrete a shell softener and drill a nice hole right into the bivalve's shell. In addition to the live prey that they eat, whelks comb the beach seeking out dead and decaying fish to eat. This makes them part of God's cleanup crew.

Whelks lay their eggs in shallow water, gluing them together in a large mass that looks like a spongy kind of paper. These papery, spongy masses, called **sea wash balls**, often wash up on the beach. People who find them sometimes don't realize that sea wash balls used to contain the eggs of an animal that creates the kind of shell for which they are looking! In

Inside each ridge of this whelk egg case, a baby whelk developed.

each disk-shaped link, a new whelk grew, shell and all.

# Winkles or Periwinkles

**Winkles**, often called **periwinkles**, produce a shell that is commonly found on the shore. Shaped like an ice cream cone, spiraling around and around the tip, a winkle shell looks a lot like a miniature version of the whelk shell. These shells are usually the size of a fingernail, but they can be

These winkles are holding onto the rocks, waiting for the tide to come back.

as large as your hand. Winkles inhabit tide pools, so you can often find one of these shells with the sea snail still inside. Surprisingly, winkles can actually survive in freshwater for several days. Most marine gastropods cannot do that.

The winkle is an herbivore. It uses its radula to scrape algae off rocks and plants. In fact, when there are a lot of winkles in a rocky tide pool, the constant scraping of all those radulas can actually increase the depth of the tide pool by as much as half an inch every sixteen years.

# Moon Snail

Another gastropod shell that is easy to find on the beach comes from the moon snail. These shells most resemble those made by snails that live on land. Interestingly, a moon snail can make its body much larger than its shell by sticking its body out of the shell and inflating its tissues with sea water.

If you have ever found a moon snail, perhaps you weren't aware that this shell belonged to a fierce ocean predator. The moon snail stalks its victim (usually a bivalve or another gastropod) and crawls up on top of it, expanding its extremely large foot so wide that it actually engulfs the entire animal, shell and all. This suffocates the prey. If the moon snail cannot suffocate its prey, it can drill into the prey's shell like the whelk can.

Moon snail shells strongly resemble those made by snails that live on land.

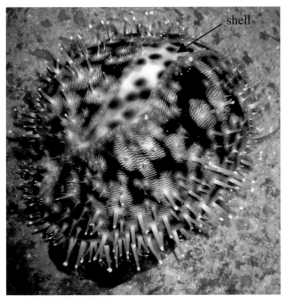

shell

You can just see a bit of the actual shell of this tiger cowry. Most of what you see is the mantle.

# Cowries

Shells made by the **cowry** have a shiny, smooth outer surface that looks like it has been polished. This is because when it is alive, the cowry's mantle actually covers its shell, keeping the shell nice and shiny. Because of this, the cowry shell appears to be a different color when the animal is alive than when it is dead. Once the animal dies, the mantle comes off the top of the shell, exposing its true color.

Interestingly enough, a cowry can retract its mantle into its shell when it feels threatened. When that happens, the cowry changes colors quickly, because retracting the mantle reveals the shell below. This can confuse predators.

For thousands of years, certain people used cowry shells as money. They have also been used to make jewelry and ornaments for clothing. If you're looking for a living cowry, you'll have to search coral reefs at night, for cowries are nocturnal creatures that feed on algae or sponges found on coral reefs.

This is the shell of a dead cowry.

# Wentletraps

Wentletrap shells are spiral with vertical ridges.

If you went to Holland and entered a grand house with a staircase that spiraled upwards, you would be seeing a **wentletrap** (went' uhl trap). Wentletraps are gastropods that are named after the Dutch word for "spiral staircase." Why? Because the shells they make are spiraled, just like a spiral staircase. These beautifully rounded, shiny shells all have distinct vertical ridges that make them look quite fancy.

Wentletraps are found near sea anemones and corals, because that's what they eat. A wentletrap attaches itself to the anemone or coral with its proboscis and then uses its radula to scrape off fragments of the animal, which it then eats.

# Cone Shells

One of the craftiest predators in the sea is the **cone shell**. Most cone shells have a similar shape, though they come in a variety of colors and patterns. This mollusk has a cone-shaped shell with a long, slitted lip that extends from the opening all the way to the top of the shell. The shell can either be smooth or have spiral ornaments.

Some cone shells are actually dangerous to people. They have a poisonous, harpoon-like proboscis, which they use to stab their prey. The tip of the proboscis injects poison into the victim, which paralyzes and then kills the unfortunate creature. This powerful and potent poison is used on both prey and predators. If a person steps on one of these cone shells, the animal will think that the person is a predator and will use its poison.

The cone shell prefers to eat fish, other mollusks, or sea worms. Once it has stabbed its prey and injected the poison, the cone shell expands its stomach, much like a snake would swallow prey, and engulfs the entire animal with its mouth. The cone shells that feed on worms are the most numerous, and their venom is usually not deadly to people. However, the venom of fish-eating cone shells found off the coast of Australia can kill a man quickly. Someone finding one of these cone shells may pick up the interesting creature to examine it, feeling only a small prick in his or her skin. Though barely felt, that small prick injects strong poison into the person's body. If not treated quickly, the person will die.

At one end of the cone shell, the mollusk extends out a siphon tube for breathing. Most people think this siphon tube is its dangerous proboscis, but the proboscis is found near the center of the shell. The cone shell mollusks found in North America are not deadly to people, as they aren't equipped to feed on fish and other large animals.

*In your own words, describe what you have learned about conchology so far. Be sure to include information about conchs, whelks, winkles, cowries, and cone shells.*

The structure extending out of this cone shell is the siphon that the animal uses to take in water so that its gills can get oxygen.

# Limpets

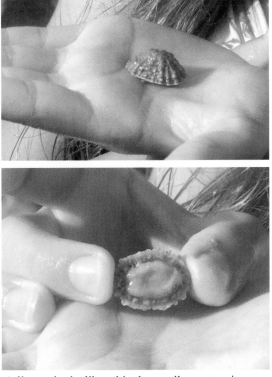

A limpet looks like a bivalve until you turn it over and see that it has only one valve to its shell.

When you see a **limpet**, you may think you are seeing a bivalve, because from the top, it looks a lot like a clam. These animals create shells that look like little oval saucers with a slight cone shape, often with a hole right in the very top. The hole is used to expel waste.

The shape may remind you of half a bivalve, but a limpet is really a gastropod, because it has only one valve to its shell. You can see this by turning a limpet over.

Dozens of limpets can be found attached to rocks and piers near the shore when the tide goes out. Because they suction themselves to these rocks with amazing force, you'll have trouble prying them loose. Even in pounding waves, a limpet holds on tightly. When the tide is in and the limpet is safely covered with water, it slides around eating algae and other plants. When the tide goes back out, the limpet goes back to its spot in the rocks. Amazingly enough, in some species, the limpet returns to the exact same spot it was on before, which it recognizes by a little indentation it made in the rock with its radula.  It seems that the limpet is not satisfied until it can find the indentation it made with its own radula.

# Abalones

**Abalones** (ab'uh loh' neez) grow to be much larger than limpets, but they behave much like limpets, clamping down on rocks and other surfaces in order to protect themselves. Unlike limpets, most abalones don't stay near the shore; they prefer the deeper waters of the Pacific Ocean, where they move about eating algae.

The large holes in this shell help to identify it as an abalone shell.

One very key feature that separates abalone shells from other animals is that their shells are equipped with several holes.  These run along the edge of the shell.  These holes are used by the abalone to expel waste, much like the one hole in the top of a limpet.  So, if you happen to find a grayish shell with a row of holes along one edge, it probably belonged to an abalone.

Abalones are hunted by man for food and for their shells.  They are considered a delicacy, especially on the West Coast of the United States.  You may have seen an abalone shell for sale, because the inside lining, called **mother of pearl**, is shiny and beautiful.  All mollusks have mother of pearl linings in their shells, but the mother of pearl that comes from abalones is particularly prized.  Because of this, the insides of abalone shells are often used to make jewelry.

# Slipper Shell

Like a slipper, this slipper shell has a shelf that covers part of the shell.

The slipper shell is produced by another gastropod.  The reason it's called a slipper shell is that if you hold the shell upside down, there is a little shelf that covers the upper tip of the shell, much the way a slipper covers the upper tip of a person's foot.  However, since this snail only grows to be 2 inches long, you could never use its shell for a slipper!

The slipper shell is a filter feeder that feeds on phytoplankton in the water.  These little creatures live out their lives attached to hard surfaces, including the backs of other snails, sea scallops, horseshoe crabs, and even one another. You'll often find them stacked one on top of the other.

Like many gastropods, slipper shells are hermaphrodites. They start out life as boys, but some become girls as they get older. Unlike most gastropods, however, a mother slipper shell keeps her eggs under her shell until they hatch. Each of the young then travels about looking for the perfect place to settle down for the rest of its life.

*Explain some of the interesting things you have learned about slipper shells, abalones, and limpets.*

This empty moon shell is now host to several creatures: a hermit crab, barnacles, and three slipper shells. Can you find the three slipper shells?

# Nudibranchs

gills

A nudibranch carries its gills on its back.

Study the beautiful blob to the left. Is it cabbage? Is it someone's hat that fell into the ocean? No. What could that thing be? Why, it's a **nudibranch** (noo' dih bronk)! The name "nudibranch" comes from two words (one Latin and one Greek) that mean "naked gills" and is very descriptive of this group of gastropods. They are some of the most beautiful animals in the ocean, with colorful, frilly gills that come in many shapes and sizes. Nudibranchs come in many colors, and some are even **bioluminescent** (by' oh loo' muh nes' uhnt), which means they glow in the dark!

These colorful clowns of the sea don't perform, but they do eat. And boy, do they eat a lot! They eat soft corals, sponges, and sea anemones. You've probably heard the phrase, "You are what you eat." Some nudibranchs get their *colors* from the things they eat. Yellow nudibranchs, for example, tend to eat yellow sponges and sea anemones. In addition, when some nudibranchs eat a sea anemone, they can actually use the sea anemone's weapons as their own. Do you remember that sea anemones have stingers that can paralyze prey? Well, certain nudibranchs can take those stingers from the sea anemones they eat and use them to protect themselves!

With or without an anemone diet, nudibranchs are usually bad to eat, and their bright colors advertise to predators that they taste disgusting. Though many nudibranchs are benthic, crawling along the ocean floor, one type oozes out bubbles, which lift it up onto the surface of the ocean where it dines on plankton such as larvae and jellyfish. Some, like the Spanish dancer, can also swim.

Although it looks a bit like a nudibranch, this is really a marine flatworm.

Some creatures, like certain marine flatworms, take advantage of the fact that nudibranchs are bad to eat. They mimic the nudibranchs, matching their colors and patterns. It is believed that this mimicry (mih' mih kree) keeps the flatworms safe from predators that would never eat a nudibranch but would happily gobble down a flatworm.

# What Do You Remember?

What is the main difference between bivalves and gastropods? How do bivalves filter feed and breathe? Where are live clams found on shore? How can you tell the age of a clam? Which bivalves cling to rocks and other surfaces? How and when do they find food? Where are pearls found and how are they formed? How do scallops swim? What does the term "gastropod" mean? What is a radula? What is an operculum? What kind of gastropod makes a shell that has a wide, pearly colored lip that flares outward? What kind of gastropod has a shell with several large holes? What kind of gastropod has no shell? What do some nudibranchs do with the stingers of the sea anemones that they eat?

# Notebook Activities

After you have recorded all the fascinating facts you want to remember for your notebook, make a separate page with information about bivalves and gastropods. Illustrate each animal and write down some of the interesting things you have learned about them. The *Zoology 2 Notebooking Journal* has pages for these activities.

**Older students:** Make a shell field guide, drawing illustrations of and writing down information about many of the shells you might find on the beach.

# Ocean Box

Now you need to add some mollusks to your ocean box. If you can find them, use real shells to make your mollusks. You can stick a piece of clay underneath a snail shell to make it look like a sea snail still lives in the shell. You can create nudibranchs by mixing bright colors of clay together.

# Experiment
## *Resonance*

**You will need:**

♦ Several bottles of different sizes

Have you ever put your ear up to a seashell, especially a large gastropod shell? If you have, you probably heard a gentle humming sound coming from inside the shell. Some say that you are hearing the sound of the ocean. Of course, that can't be true. After all, the shell is no longer in the ocean. It is in your hand. It cannot "store" the sound that the ocean makes.

What you are really hearing is the air inside the shell vibrating. Why is it vibrating? Well, the molecules in the air around the shell are bouncing against the outside of the shell, making the shell vibrate. The vibrations on the outside cause the air inside the shell to start vibrating as well. We call this **resonance** (rez' uh nuns). You can make your own resonance using bottles instead of seashells.

1. Remove the lids from all of your bottles.
2. One by one, blow across the mouth of each bottle. If you work at it, you should be able to produce a constant sound.
3. Compare the sound that one bottle makes to the sound that another bottle makes.
4. Notice how the different bottles produce higher or lower sounds, depending on the size and shape.

What does this experiment tell you? When you blew into the bottle, you were creating a resonance in the bottle. Air began vibrating in response to what you were doing. Did you notice that the size and shape of the bottle affected what kind of sound the resonance produced? It's the same with seashells. If you compare the sound you hear in one seashell to that of another seashell, you will discover that the size and shape of the shell affects the sound that you hear.  Next time you are at the beach, try placing different shells up to your ear and noticing the different sounds they produce.

# Project
## *Make a Conchology Box*

Would you like to become a conchologist?  It's a great hobby that can last a lifetime!  You can begin by getting a good shell field guide and then creating a box in which to keep your shells.  Some conchologists keep their shells in small plastic boxes with lids, but you can begin by using shoe boxes and creating slots in them.

**You will need**:

♦   A least one shoe box

♦   Scissors

♦   Glue

♦   Some cardboard (You can cut up cereal boxes or other thin cardboard boxes.)

♦   Cotton batting or a sheet of foam

1.  Open one of your shoe boxes.
2.  You will want to make compartments in the box so that you can separate the seashells you collect from one another. You can do this by first cutting strips of cardboard that are as tall as the box.
3.  Take one strip and bend both ends to make flaps.
4.  Glue one of the flaps to the side of the box.  Then, glue the other side to another strip of cardboard that will form the wall of the next compartment.
5.  Continue to do this until you have several compartments of different sizes, as is shown in the picture on the right.
6.  When you are done, you can spray paint your box if you wish, or just leave it the way it is.
7.  After you have prepared as many compartments as you like, lay down cotton batting in each compartment.  You could also use foam that is cut to fit each compartment.
8.  Now that your box is ready, collect some shells to put in it!  When you take a trip to the beach, wake up early in the morning to find seashells on the beach.  Beachcombing is a sport where the early bird gets the worm.  Even if you can't get to the beach very often, you can buy some shells to start your collection.  The course website mentioned in the introduction to this book has links to places where you can buy shells.
9.  As your collection grows, you will need more boxes like this.

# Lesson 10
# Cephalopods

By now, you've learned that God created some truly incredible creatures, but what if I told you that there are animals that can change color in a split second and even change the texture of their skin? Are you impressed? What if I told you they can also change the shape of their bodies to fit into small cracks and crevices? In addition, suppose I told you they have three hearts that pump *blue* blood, eyes that can see better than ours (in some ways), the ability to make their own ink, and are powered by jet propulsion? Would you believe that such animals really exist? Well, they

Cephalopods like this octopus are amazing animals.

do! They have representatives in every part of earth's oceans, from the tropics to the poles and from the shore to the abyss. They have inspired legends and stories throughout history and are thought to be the most intelligent invertebrates in creation. **Cephalopods** (sef' uh loh podz) are all this and more. Let's investigate these amazing animals, shall we?

The **octopus**, the **squid**, the **nautilus**, and the **cuttlefish** are all cephalopods. These animals belong to class **Cephalopoda** (sef' uh loh poh' dah), which is part of phylum Mollusca. In other words, cephalopods are mollusks. Yes, they are mollusks, just like the gastropods and bivalves we learned about in the previous lesson. Do you remember that "gastropod" means "stomach foot?" Well, "cephal" refers to the head, and, of course, you already know that "pod" means "foot." So cephalopods are "head foot" animals. They consist of a head with a bunch of tentacles attached. Although they are mollusks, cephalopods are different from the gastropods and bivalves we studied in the previous lesson. For example, most bivalves and gastropods don't have much of a head or brain. Cephalopods, on the other hand, have well developed senses and large brains, making them really smart for mollusks. Also, unlike gastropods and bivalves, cephalopods are nektonic animals. Do you remember what nektonic animals are? I hope so!

Now, here's something interesting; we call the tentacles of a cephalopod "arms," not feet or legs. An octopus, then, has eight arms, not eight legs. That's a bit strange, since they are considered head-*footed* animals, not head-*armed* animals. But that's the way it is. Squid and cuttlefish have two more arms, for a total of ten. These two extra arms are really tentacles, but they are still called arms. Isn't that strange?

Cephalopods eat mostly fishes and crabs. They have suction cups on their arms that they use to capture their prey, and then they bite it into pieces with their beaks. Yes, in the center of their bodies, under all those waving arms, cephalopods have beaks! Some cephalopods can use their beaks to inject a poison into their prey when they take a bite. Often, the bite is so venomous that the prey is quickly killed by it.

Can you find the octopus in this photograph? It is so well camouflaged, it is hard to see. Check the course website if you can't find it.

One special gift God gave cephalopods is the ability to camouflage themselves very quickly. Many animals (lizards, for example) can change their colors in order to camouflage themselves, but this process generally takes a while. Well, that's not the case with these chameleons of the sea. Cephalopods can camouflage themselves in a split second! This is a great form of defense, and cephalopods really need it. You see, despite the fact that they are mollusks, they don't have a shell covering the body. As a result, cephalopods are vulnerable to predators. To make up for this, God equipped them with cells called **chromatophores** (kroh mat' uh forz). These cells have pigments (chemicals that produce color) which can be manipulated to change the color of the skin, allowing them to blend in with their surroundings.

Most cephalopods also carry around ink. No, the ink isn't used to write letters; the ink is a thick, dark liquid that they can shoot out when they are in danger. This big cloud of black, inky liquid completely darkens the waters surrounding the cephalopod so that a predator is unable to see it. Since the predator can't see the cephalopod, it is able to escape unharmed.

Have you ever considered your eyes a good form of defense? Well, if you were a mollusk, a good set of eyes would be a great thing to have. You could see what was approaching so that you could get inside your shell (if you have one) or make a quick exit. Yes, indeed, eyes are a great defense that some mollusks don't have. However, cephalopods have extremely well developed eyes and can detect predators from a long distance away. This is another defense God gave them for their protection and survival.

Of course, their excellent eyesight also helps cephalopods find food. In addition to this, they have special senses in their arms to locate and choose the best food. They can actually taste with their suction cups!

# Propulsion

Since cephalopods are nektonic, we know that they can swim, but how do they do this without fins or flippers?  Well, God designed the cephalopod to allow water inside its body and then quickly squeeze the water out, sending a jet stream out of its body, propelling it backward.  This is a form of **jet propulsion** (pruh pul' shun).  The stream of water goes out a tube called the **hyponome** (hi' puh nohm).  To go even faster, some cephalopods can tighten their bodies into a streamlined, bullet shape so that they glide easily through the ocean.  The cephalopod can change whatever direction it wants to go by pointing the hyponome in the opposite direction.

# Cuttlefish

The **cuttlefish** (kut' uhl fish) is a graceful animal that lives on the bottom of the ocean, usually lying flat like a sting ray.  To stir up the creatures hiding under the sand, it blows a jet of water out its hyponome, upsetting the sand and uncovering the creatures that live there.  Other times, it surprises its prey.  Lying completely camouflaged and unseen by the other sea creatures, it suddenly darts out, ambushing its prey, which includes fishes, crabs, mollusks, and shrimp.  With one bite of its beak, it injects

Although odd-looking, the cuttlefish is a graceful animal.

poison into its prey, and its helpless dinner has lost the battle.  Like many other cephalopods, most cuttlefish are nocturnal, which means they are nighttime hunters.

Believe it or not, the cuttlefish has a shell.  However, like most cephalopods that have one, the cuttlefish's shell is on the *inside* of its body, not the outside.  As a result, we say that it has an **internal shell**.  This internal shell is often called a **cuttlebone**, even though it is a shell, not a bone.  The cuttlefish has a flattened body and a fin running all around the body.  It has ten arms – eight that have suckers lining the bottom and two tentacles that are stored in pouches under its eyes.  These tentacles can shoot out of the pouches to capture prey.

The cuttlefish's skin is usually mottled brown, red, and black, but it can quickly change its color and even its texture to match its surroundings.  This happens so fast that it is actually hard to see it happening.

# Squids

A **squid** has ten arms – eight regular ones and two super long feeding arms that are used to grab prey and bring it to the other arms.  It holds its prey with its eight arms and uses its beak to bite the prey into smaller pieces it can swallow.  A squid typically feeds on fishes, shrimp, crabs, and other squids.

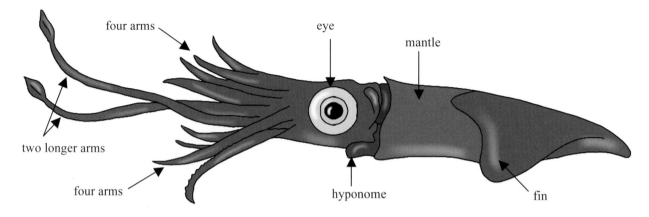

four arms

eye

mantle

two longer arms

four arms

hyponome

fin

The cone-shaped end of the squid is covered in its mantle, and at the end of the mantle, there is typically a fin.  This often makes the body of a squid look like an arrow.  This arrow shape is useful when swimming in the open ocean, where squids prefer to stay.  The fin is not used to push the squid through the water, however.  It is a stabilizing fin that helps the squid hold its body the way it wants to. When threatened, a squid can use its jet propulsion to swim away, and if necessary, release its ink to confuse the predator.

Not only are squids eaten by sharks, many larger fishes, and even other squids, they are also eaten by people.  In fact, Asian people eat squid almost as frequently as people in the United States eat chicken.  In Asian countries, squids are caught by the thousands each day and hung to dry on racks.  Squid can often be found on restaurant menus.  It is usually called calamari (kah' luh mah' ree).

Squid drying on racks is a common sight near the shore of many Asian countries.

Fishermen are able to catch many squids at once because they often swim in groups called **shoals** that contain huge numbers of individuals.  This can lead to some amazing sights.  For example, some squids can actually leap out of the water to avoid predators.  Pacific flying squids, for example, have been seen leaping many feet out of the water to avoid a predator that has come upon the shoal. The sight of so many squids leaping out of the water to avoid the predator can be breathtaking!

Many squid are bioluminescent.  Some use this God-given ability to hide from predators.  If a squid is swimming and emits a faint glow, it can blend in with the light coming from above.  This is called "countershading," and can make the squid nearly invisible to predators that are below it.

# Reproduction

A mass of squid egg cases often resembles a flower.

After a male and female mate, the female squid lays eggs.  The eggs are laid inside an egg case, and since the squid is most likely a part of a shoal, it is laid with many other egg cases from many other mother squids.  The egg cases are usually anchored to the sea floor.  Since squids travel in large shoals, many squids are laying eggs all at once.  They tend to anchor their egg cases very close to one another, so you usually find squid egg cases in clumps, and those clumps often resemble a flower.

Often, the male will die shortly after mating, and the female will die once she has released her eggs.  As a result, most squid reproduce only once.  Like most cephalopods, squids do not live very long.  Although there are some long-lived species, most squids live for only one or two years.

When the eggs hatch (usually within a few months), little miniature squids emerge, looking identical to the adults except that they are tiny.  Although egg cases do not seem appealing to most predators, as soon as they hatch, many predators enjoy feeding on the baby squids.

# Giant Squids

For many years, people would hear tales from fishermen about **giant squids** out at sea. According to the stories, an enormous squid would wrap its arms around a whale, causing a terrible battle to ensue.  Whalers claimed that sometimes, when they caught a whale, it would have giant round scars as big as dinner plates on its body. Other times, when they cut open the stomach of one of their whales, they would find squid arms as long as 30 feet with suckers up to four inches wide. The whalers who told these stories either chopped

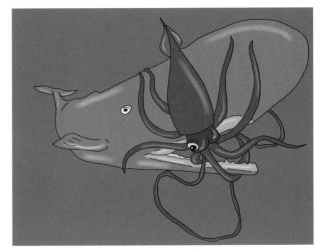

up the squid parts to eat or use as bait, or they threw them back out to sea before scientists were ever able to examine them. However, in 1861, a French ship was able to bring back parts of a giant squid so that scientists could study them.

Then, in the late 1800s, several giant squids were washed up on shore, and it was clear that giant squids really do exist. Since then, many giant squids have washed ashore or have been found dead out at sea. Scientists have been increasingly fascinated with these mysterious creatures, but few have ever seen them alive. Many methods of finding and filming them have been tried, but most have failed. Because very few people have ever actually seen or filmed a living giant squid, they are one of the great mysteries of the deep. Scientists believe that giant squids cannot survive for long in warm water and that this is the reason it is difficult to find a living one. As a giant squid gets closer to the surface, the water gets too warm for it to survive for long. Thus, scientists think that they spend most of their time in the deep, cold sea, where few can venture.

This dead giant squid that washed up on shore is about 15 feet long.

Even though it is difficult to see them alive, studying dead giant squids has taught us a lot. It is thought that male giant squids can reach lengths of more than 30 feet, while the females can grow to more than 40 feet long. They have *huge* eyes that can be up to a foot across. That's bigger than a basketball! Giant squids seem to have the same basic features as their smaller counterparts, they are just a lot bigger.

Now that we know giant squids exist, scientists have examined whales closely and have found some with large, round scars. They look suspiciously like the marks the suckers on a giant squid would leave. This indicates that the stories of giant squids fighting with whales might be true. Most likely, the whales got those marks from fighting the squids while they were trying to eat them. A sperm whale's appetite for giant squids has aided in their study, because Japanese scientists recently used sperm whales as guide to find live giant squids. These scientists ended up capturing the first photographs of live giant squids. If you go to the course website I told you about in the introduction to this book, you will find a link to those photographs.

*Take some time now to explain all that you have learned so far about cephalopods and squid.*

# Octopuses

If the plural of cactus is cacti, wouldn't the plural of **octopus** be octopi? You would think so, but that is not correct. The word for more than one octopus is "octopuses." Textbooks sometimes call them "octopods." Very few people use that term, however. You'll often read "octopi" as the plural of octopus, but it is not really correct. In English, the correct word for more than one octopus is octopuses.

While squids swim in the open waters, octopuses like to live in shallow waters, spending most of their time on the sea floor. Most are nocturnal, hiding under plants or in any little crevice they can find during the day and emerging at night to feed. Their bodies are very pliable; a large octopus can squeeze into a very small slit. Though their natural color is usually a grayish brown or an orange-tinged color, their ability to quickly change colors and blend into their surroundings makes them hard to spot. Although most octopuses you will run across are small, they can grow to be as big as thirty feet across!

Octopuses like to live in shallow waters.

Since "octa" means "eight," it should not surprise you that octopuses have eight arms. Each of these arms is lined with a double row of muscular suckers that can be used to pull the animal along the rocks, reefs, and ocean floor while it looks for food. One kind of octopus even crawls up onto land to

look for food on the shore. An octopus also uses its suckers to fasten onto its prey, which includes shrimp, crabs, and other crustaceans. It then easily bites into the captured animal with its strong beak.

When emerging at night to look for food, an octopus will often keep one or two arms attached to a rock while the rest of its body floats about looking for food. If it senses danger, the arms that are anchored to the rock are used to quickly pull the octopus back down into its hiding spot.

Note the double rows of suckers on these octopus arms.

Most octopuses are safe to handle, but the blue-ringed octopus that lives off the coast of Australia is so poisonous that it is a danger to people. A blue-ringed octopus is a small creature that hides in shallow waters and rocky pools near the shore. It is normally dark brown to dark yellow in color, but when it is frightened, it turns vivid yellow with bright blue rings. One of these small octopuses has enough poison to kill ten people! Strangely, the octopus's bite doesn't really hurt, but the octopus's poison quickly flows into the small, mostly painless wound. If the victim doesn't get to a hospital soon, the bite could very well be fatal.

# Feeling Colors

Some octopuses can use their chromatophores for more than camouflage. If they are not blending in with their surroundings, they often express themselves using color. When they are relaxed, for example, they're a dull, grayish brown or orange-tinged color. When they become angry, however, they may turn red. On the other hand, if they feel scared, they can turn white. Can you imagine what it would be like if your parents, brothers, sisters, and friends could turn different colors depending on how they feel?

Since it is mostly white, this octopus may be frightened.

# Reproduction

After a male and female octopus mate, the male usually dies. Then, the female lays eggs that are about the size of grapes, attaching them to rocks and other underwater surfaces. Like a devoted mother, she stays with them every day, cleaning them and caring for them. The entire time she is caring for her eggs, she never leaves them, even to eat or search for food. About four months later, the eggs hatch and the mother usually dies of weakness and starvation. The tiny hatched octopuses look much like their parents and can immediately fend for themselves in the big ocean world. They set out on their own to live a solitary life, as most octopuses prefer to be alone. They usually live for about two or three years. Even the large, 30-foot long octopuses only live for a few years, beginning life larger and growing faster than other octopuses.

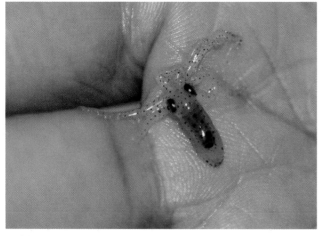

This octopus baby is so small it barely takes up any room in this man's hand. Even though it is recently hatched, you can tell it is an octopus.

# Octopus Brains

With its large eyes and giant brain, an octopus has been shown to be one of the most amazing creatures God created. It is remarkably intelligent. Using problem-solving skills, octopuses have been known to get out of their aquariums or climb aboard ships to open buckets of crabs. In laboratory experiments, octopuses have been able to open jars. They have also been seen in aquariums playing with objects!  Yes playing!  Scientists have watched while several octopuses played with a pill bottle that was tossed into the aquarium.  The octopuses squirted water out of their hyponomes, making the bottle fly into the curricular current of water, going around in a circle and back as it was squirted again and again.  Some of the octopuses were found to play with the pill bottle for nearly 30 minutes!  Because of their large brain and obvious intelligence, it's believed that they feel pain in the same way we do, so scientists are not allowed to perform surgeries or other painful procedures on them without using anesthesia.

# Seeing Eye to Eye

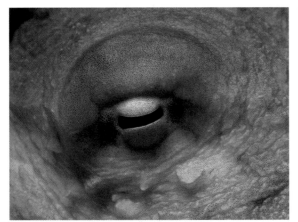
Octopus eyes are different from human eyes.

One very fascinating aspect of the octopus is that its eyes are very complex.  In fact, the octopus doesn't have a blind spot.  "What's a blind spot?" you ask.  Well, each of your eyes has a certain place on it that is unable to see.  As a result, when you are looking in a particular direction, part of what is in that direction will not be visible, because its image falls on your eye's blind spot.  An octopus, however, does not have this problem.  The reason this is the case has to do with how the eye is "wired."

In a human eye, the parts that detect light (called the **photoreceptors**) are actually behind the nerves that send what these photoreceptors see to the brain.  As a result, light must travel *through* the nerves before it can be detected by the eye.  Well, in the spot where the nerves bundle up and go to the brain, there is no room for any photoreceptors.  Because of that, the eye cannot see any light hitting that particular spot.  In the octopus, the nerves are *behind* the photoreceptors.  As a result, the nerves bundle up and go to the brain *behind* the photoreceptors.  That way, there is no spot on the eye that doesn't have photoreceptors.

Why do the eyes on people have the nerves in front of the photoreceptors when the eyes on octopuses have them behind the photoreceptors?  Mostly, this is because octopuses are not exposed to the same kind of light that people are exposed to.  As a result, their photoreceptors can receive *all* of the light that hits them.  Compared to animals under water, people are exposed to higher levels of ultraviolet light, which can cause problems such as cancer.  The nerves in front of the photoreceptors filter this ultraviolet light, protecting the photoreceptors from harm.  The "wiring" of the human eye, then, allows us to see well *above* water, and the "wiring" of an octopus's eye lets the octopus see well *below* water.

Another difference between human eyes and octopus eyes is found in the lens.  The eye's lens focuses light so that we can see what we want to see sharply.  The lens in a human eye is flexible.  It changes shape in order to change focus.  The lens in an octopus eye, however, is stiff and not very flexible at all.  Instead, the lens is movable.  It moves back and forth within the eye in order to focus.  This is the way a camera focuses.  When a camera focuses on an object in order to take a picture, the lens moves back and forth until the image that the camera sees is in sharp focus.

## Try This!

To test your blind spot, do the following exercise:  Hold the book far from your face.  Cover your right eye.  Now look at the plus symbol with your left eye.  Keeping your left eye focused on the plus symbol, slowly move the book towards your face until you notice the red dot disappears from view.  When the red dot disappears from view, its image is hitting the blind spot on your eye.  If you continue to move the book closer to your face, you will see the red dot reappear again, because its image has moved away from the blind spot on your eye.  If you were an octopus, you would never see the red dot disappear!

# Nautilus

If you were scuba diving at night in the South Pacific or Indian Ocean and happened to see a large, cream-colored shell with brown wavy lines across it float past, it was probably the cephalopod with the most arms: the **chambered nautilus** (not' uhl us).  Not only does it have more arms than any other cephalopod, is the only one that creates an external shell.

As soon as a nautilus hatches, two of its arms start to make a thin shell.  The body of the nautilus doesn't fit comfortably inside the shell; its eyes and most of its arms stick out of the shell most of the time.  It will only completely retreat into its shell when it feels threatened.  It's not that the nautilus' body is so big, it's because the shell is divided into a bunch of different compartments, and the nautilus only lives in the outermost compartment.  Each compartment is a gas-filled chamber connected to the other chambers by a small tube.  Each year, the nautilus grows and creates a

The arms, eye, and hood of this nautilus are visible.

new, larger chamber in which to live.  Its body occupies only the most recently created chamber.

A nautilus has a thickened area over the head, which is called the **hood**.  When the animal withdraws into its shell, the hood covers the opening to the shell.  In other words, it is a protective "lid."  Can you see the arms sticking out of the shell in the picture above?  Guess how many arms the nautilus has.  Octopuses have eight; squid and cuttlefish have ten.  Unbelievably, a nautilus can have anywhere between 38 and 90 arms!

The nautilus controls its depth using the tube that connects the gas-filled chambers in its shell.  If it sucks in gas through the tube, water enters the chambers, making the nautilus heavier.  This makes the nautilus sink.  If it forces gas out of the tube, that pushes water out of the shell, making the nautilus lighter.  This makes the nautilus rise in the water.  You will do an experiment at the end of this lesson that shows you how this works.  In addition to rising and sinking, the nautilus swims using the same jet propulsion system that other cephalopods use.

During the day, the nautilus stays hidden in crevices in the bottom of the ocean, more than 1,000 feet below the surface of the water.  At night, it fills its chambers with gas to rise higher looking for food.  Its favorite thing to eat is crustaceans.

*We have completed our study of cephalopods.  Before we move on to chitons, tell someone all that you remember about octopuses and nautiluses.*

# Chitons

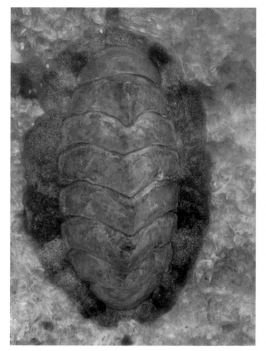

The eight plates that make up this chiton's shell are easy to see.

Have you ever picked up a little creature out in your yard called a pill bug?  Sometimes, it is called a doodlebug or a rolypoly.  These arthropods are gray and can roll into a tiny ball.  The last mollusk that we'll study has some things in common with the pill bug, but it is really very different.  This mollusk is not a cephalopod, gastropod, bivalve, or nudibranch.  It is called a **chiton**.  A chiton is a bit like a gastropod, because it has a single shell that suctions to rocks, similar to a limpet.  However, the chiton's shell looks more like a pill bug or a crustacean, with hard plates that fit tightly together, overlapping one another, and forming an oval-shaped shell.

With neither eyes nor arms, chitons are in their own class, called **polyplacophora** (poh lee plah' koh for' uh), which means "many plates."  Can you see why it is called this?  The plates, of course, refer to its shell.  Count the plates on the chiton pictured to the left.  How many do you see?  A chiton has eight plates, and underneath the plates is a foot that moves the chiton over rocks and other structures, both in and out of the water.  It also has a radula.  Some use it to scrape algae off rocks, while others are carnivorous, catching tiny zooplankton and other small animals that live in the shallows.

Chitons can be less than an inch long or up to a foot long. They come in several different colors: black, red, pink, and even blue. The shell might be shiny or dull, depending on the kind of chiton it is.

If you picked up a chiton and turned it over, which would be extremely difficult since they use suction to cling tightly to rocks, you would see no eyes, legs, or arms. All you would see is a wide foot, equipped with a radula that it uses for scraping surfaces. Even if you were able to pry the chiton loose, you might not get a chance to see the foot or radula, because many chitons can roll into a ball, just like a pill bug! Its mantle is right under the shell, with the lowest portion of the mantle coming out under the edge of the shell to help its foot grip the surfaces upon which it clings or slides along. This protruding portion of the mantle is called the **girdle**. As with some of the gastropods we already discussed, like the moon snail, the girdle may extend out over the bottom portion of the shell covering the bottom half of the chiton. Depending on the species, the girdle may be smooth, hairy, scaly, or spiny, giving different chitons different looks.

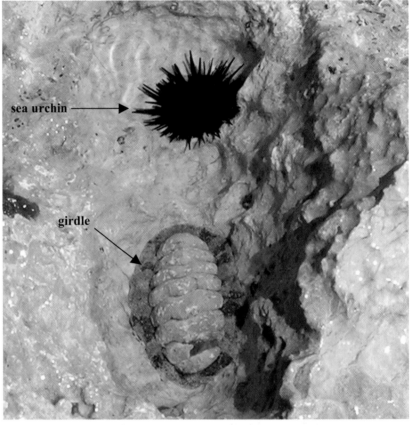

This chiton is sharing a tide pool with a sea urchin. Both cling so strongly to the rock that it is very hard to remove them.

Most chitons are nocturnal, feeding at night, and hiding under rock ledges during the day. However, they are easy to find in tide pools, as they don't hide in cracks and crevices or camouflage themselves the way octopuses do.

# What Do You Remember?

What does "cephalopod" mean? What are the four different kinds of animals in the cephalopod group? How do cephalopods swim? What do cephalopods usually eat? What kind of mouth do cephalopods have? What are some of the defenses that cephalopods have? What is the internal shell of the cuttlefish called? How many arms do cuttlefish and squids have? What does a squid usually do after it mates or lays eggs? How many arms does an octopus have? Why do scientists think octopuses are intelligent?  What is different about the nautilus compared to other cephalopods?  How does the nautilus move up and down in the water?  Describe a chiton.  Which land animal is it like and why?  How is it like a gastropod?  Where might you find a chiton during the day?

# Notebook Activities

Make five notebook pages, one for each of the animals you learned about in this lesson: cuttlefish, squid, octopus, nautilus, and chiton.  Make an illustration and write down the interesting things you have learned for each of them.

**Older Students:** Write a story that describes a sperm whale hunting and attacking a giant squid.  Assume that the giant squid fights back when the sperm whale attacks it, and write about the battle between these two giants.

# Ocean Box

Create some cephalopods for your ocean box.  Make at least one squid and one octopus.  If you want, buy some "google eyes" at a hobby shop to use for their eyes.

# Experiment

We are going to create our own little cephalopod and learn about buoyancy.  You will place a small medicine dropper in a plastic bottle.  You will then conduct an experiment to make the medicine dropper sink and float by displacing the amount of water or air in the dropper.  This is similar to how some cephalopods (like the nautilus) are able to move up and down in the water.

**You will need:**
- A Scientific Speculation Sheet
- A small eyedropper
- One empty, small, clear water bottle or soda pop bottle with a lid
- A glass of water

**Optional items to make your dropper look like a cephalopod:**
- A google eye
- 1 chenille stick (pipe cleaner) or yarn
- Double sided tape

**Optional:**
1.  To decorate your medicine dropper, cut the chenille stick in half or cut the yarn into pieces.
2.  Wrap the chenille stick or yarn pieces gently around the top of the medicine dropper, being careful not to squeeze the dropper.
3.  Tape the google eye on the dropper.

**To do the experiment:**

1.  Fill the empty plastic bottle with water.
2.  Fill the water glass with water
3.  Place the end of the medicine dropper in the glass of water.
4.  Get some water inside the dropper by squeezing the rubber bulb while the end is in the water.
5.  Now release the dropper and see if it floats upright in the water. If it sinks to the bottom of the glass, you need to remove some water. If it does not float upright, you need to get some more water into the eyedropper.
6.  Continue to work with the amount of water in the eyedropper until it just barely floats upright in the water.

7.  Now, place the eyedropper in the bottle and screw on the cap tightly.
8.  You are going to squeeze the bottle and see what happens to the eyedropper. Before you actually do that, however, write down a hypothesis on your Scientific Speculation Sheet. What will happen to the eyedropper when you squeeze the bottle?
9.  Now, squeeze the bottle and see what happens.
10. Release the bottle and see what happens.
11. Write down your observations and finishing recording the experiment on your Scientific Speculation Sheet.

What happened? Well, if it worked right, the eyedropper should have sunk when you squeezed the bottle and risen when you released it. If it didn't work, you may need to adjust the amount of water in the bottle or inside the dropper. Keep experimenting until it works.

Why did the eyedropper behave as it did? Well, as you applied pressure to the bottle by squeezing it, you also applied pressure to the inside of the eyedropper. This squeezed the air inside the eyedropper, which made more room for water. Thus, water rushed into the eyedropper, making it heavier than it was before. This caused the eyedropper to sink. When you released the pressure on the bottle, water rushed out of the eyedropper, making it lighter. This allowed the eyedropper to rise back up to the top. Some cephalopods (like the nautilus) can control their buoyancy in the same way. They can take in water to become heavier, making them sink, or they can expel water to become lighter, which makes them rise in the water. Submarines do the same thing. They fill tanks with ocean water to sink, and they push the ocean water out of the tanks in order to rise.

# Lesson 11
# Echinoderms

Although most echinoderms are harmless to people, this crown of thorns sea star has venomous spines that cut the skin. The poison is not fatal for people, but it does cause sickness and swelling.

How did the starfish and the sea urchin pay for their lunch? Can you guess? With a sand dollar! That's a classic **echinoderm** (ee kye' noh derm) joke, since all those animals are members of phylum **Echinodermata** (ee kye' noh der mah' tah). "Echinoderm" comes from two Greek words: *echino* (which means "spiny") and *derma* (which means "skin"). Thus, "echinoderm" means "spiny skin." If you have ever felt the skin of a living sand dollar or starfish, you can understand why it's called spiny skin. It has a rough, bumpy texture, with little spines found covering it. The animals in this group include sea cucumbers, feather stars, brittle stars, sea urchins, sand dollars, and starfish, which are more properly called sea stars.

All these spiny-skinned creatures have a few things in common: no eyes, no brain, and **tube feet**. These little tube feet are fascinating. They are essentially tubular strands with suction cups on the end. Echinoderms move by using their tube feet to suction onto a surface and then push or pull across it. They can also use them to hold onto a surface. Sea stars suction onto bivalves in order to force them open, and sea urchins suction onto hard surfaces to stay in one place.

God designed these little tube feet to do two other important jobs. First, they help the echinoderm breathe! In addition to special gills that stick out from the skin, the tube feet take in oxygen from the water and release carbon dioxide as waste. Second, they are like tongues, tasting everything the echinoderm touches to see what might make a good meal. Since the creature has no legs, eyes, or nose, these little tube feet are crucial to the echinoderm's survival!

You can see the suction cups on the ends of this sea star's tube feet as it clings to the side of its aquarium.

Another thing most of these little guys have in common is radial symmetry (sih' muh tree). "What is that?" you ask. Well, each of these animals (except the sea cucumber) has a mouth in the center of its body, and the rest of its body radiates out from the center.

Since they spend their time on the seafloor, can you tell me if echinoderms are part of the nekton, plankton, or benthos? They are definitely part of the benthos. However, like most benthic animals, when echinoderms are larvae, they float about in the water as part of the plankton population. Let's explore some of the different echinoderms in God's creation.

# Sea Stars

**Sea stars** are found all over the ocean, from the beaches to the continental slope. They can be as small as your pinky finger or 3 feet across — almost as big as you! Most sea stars have five arms, called rays (like the sun's rays), radiating out from a central disk. If a sea star has more than five rays, it will often have rays in multiples of five; there could be ten, fifteen, twenty, or even up to thirty rays on one sea star.

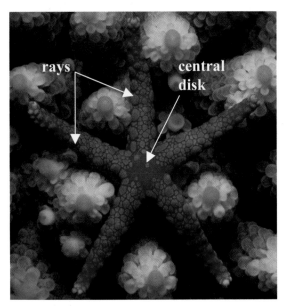

Most sea stars have five rays extending from a central disk.

A sea star's rays come from a central disk, which has a large hole on the underside. Can you guess what that hole is? If you guessed the sea star's mouth, you were right. Interestingly enough, when it eats, the sea star doesn't use its mouth the way we do. When you and I eat, we use our mouth to take in the food and chew it. Then, when we swallow, the food travels through our esophagus to our stomach. Well, in the sea star, it is completely different. The sea star's stomach actually *goes out of* the mouth to engulf the prey. Once the stomach has done that, it immediately begins digesting the helpless animal. As the food gets smaller and mushier, the sea star pulls the stomach back in through the hole to be fully digested inside. Now that's an odd way of eating, indeed!

Notice the mouth in the center of this sea star's central disk. The stomach will leave the mouth so the sea star can eat.

So what is the favorite food for this escaping stomach? Can you guess? It's a bivalve, the same food enjoyed by many creatures. The sea star crawls up onto an innocent bivalve, like a clam, and wraps its suction-footed rays around the clam. It then attaches the suction cups on its tube feet to each side of the clam and starts to pull the clam open.  At first the clam uses its muscles to hold itself closed, but eventually, the clam gets tired, and the sea star can open the shell by just a tiny crack.  Once it has done that, the sea star's stomach leaves its mouth.  Interestingly, the stomach is kind of like an octopus; it can change shape to fit into small cracks and crevices.  So, if the clam is slightly open, the sea star's stomach can slip right inside and begin digesting the tasty body of the clam. After the sea star has fully digested its

This sea star has its tube feet locked onto the clam that it has surrounded. Its rays will pull on the clam until the shell has opened enough to allow the sea star's stomach inside.

meal, wastes are eliminated from the anus, which is on the side of the body that is opposite the mouth.

Although you thought the sea star was just an interesting little treasure of the sea, you now realize how ferocious it is.  In fact, the sea star lumbers across the sand or rocks, searching for prey. Believe it or not, sea stars are some of the most numerous predators in the ocean!  When a sea star happens upon any edible creature, out comes its stomach to gobble it up.

Because sea stars feast on huge numbers of clams in the ocean, people who gather clams for a living have tried for years to get rid of them.  In order to kill the sea stars, these "clam fishermen" would catch them, slice them right in half, and toss them back into the ocean.  They believed that this would eliminate the sea stars and keep them from eating up the clams.  Do you think this was a good idea?  Why or why not?  You'll find out in a moment.

A sea star's tube feet are concentrated in grooves that run down the center of each ray.  These grooves are called **ambulacral** (am byoo lah' krahl) **grooves**.  Now remember, the tube feet help the sea star to breathe.  Interestingly enough, the oxygen that these tube feet collect is not distributed through the body by blood.  Sea stars don't have blood!  Instead, a sea star has a system of canals, called the **water vascular system**, that transports oxygen, food, and the other things it needs to different parts of the body.

Sea stars don't respond to touch very quickly.  In fact, if you pick up a sea star, you might wonder if it is alive.  Yet, by turning it over and looking at the little moving feet on the bottom, you will see that it is, indeed, a living creature.  Sea stars don't have eyes like you or me.  Instead, at the end of each arm are eye spots, which are only able to tell light from dark.  If you happen to see a sea star at an aquarium or at the beach, try to find its eye spots.  An eye spot is usually colored differently from the rest of the ray, and it generally is at the very tip.

# Making New Sea Stars

This sea star lost four of its six rays, but it is regenerating them. That's why the four rays on the right are smaller than the other two rays.

Sea stars, like many sea creatures, are able to **regenerate** parts of their bodies. What does it mean to regenerate? Well, just as crabs and lobsters are able to grow new chelipeds if they lose them, sea stars are able to grow new rays. Actually, sea stars are better at regeneration than most other creatures. Not only can a new ray grow when a ray is torn off, but if the torn-off ray has even a small piece of the central disk still attached, a whole new sea star can grow from the one ray! Sea stars are sometimes known to intentionally break off a ray and a piece of the central disk just so another new sea star can grow. Not only will the ray that was broken off grow back, but the arm that broke off will also grow into an exact copy of the original sea star. In other words, sea stars can clone themselves!

Do you remember the question I asked you about clam fishermen? They were trying to reduce the sea star population by cutting the sea stars in half and throwing them back into the ocean. Did you think that was a good idea? Well, now you know it is not. Instead of reducing the number of sea stars, they were increasing the number of sea stars, because every sea star they caught and cut in half became two sea stars!

*Explain in your own words what you have learned about sea stars.*
*Be sure to include a description of how they eat.*

# Brittle Stars

If you find a fast-moving sea star with long, thin, snake-like arms surrounding a central disk, you've not seen a sea star at all. You have seen a **brittle star**. Brittle stars are thought to be much more fragile than sea stars, which is how they got the name brittle star. It was probably named brittle star when someone picked one up and an arm fell off. At first, this makes the animal seem brittle; however, like a crab, a brittle star will drop its arm to escape if it is touched. It's not really brittle; however, it sometimes does break itself. So, if you ever see a brittle star and pick it up, it may fall

The ruler in this photo tells you that this Hawaiian brittle star is about 2 inches across.

apart in your hands on purpose!  This is a defense that God gave these little creatures to escape from predators.

A tiny brittle star was hiding under this piece of debris.

Interestingly, brittle stars do not use their tube feet for movement, but instead use muscles in their long, thin arms to scuttle rapidly about wet and dry surfaces.  They are fast movers.  This makes them very different from sea stars, as sea stars move slowly while they creep along the seafloor.

Though there are hundreds of thousands of brittle stars near the shore, you aren't likely to run across one on the beach or in a tide pool.  Why is that?  Well, during the day, they hide under anything they can find, only coming out at night to feed.  Though they are extremely numerous, they are nocturnal animals, which makes them fairly hard to find.

Now, it's important to note that not all brittle stars are the tiny, fragile stars that I have been telling you about so far.  As scientists have studied more and more sea life, they have found that there are many different kinds of brittle stars.  **Basket stars**, for example, are large brittle stars with many rays.  Although some people mistake these brittle stars for plants, they are definitely animals.  If you look at the picture of the basket star on the right, you will see that the rays form many branches so that they are almost feathery on the ends.  This is important, as basket stars are filter feeders.  They use their branched rays, which are covered with a sticky substance, to catch plankton from the water so that they can eat the plankton.  They get their name because when they stretch their rays out to catch plankton, they tend to look like the bottom of a basket.  Because they are nocturnal, they stretch out at night.  During the day, they generally sleep by curling up into a tight ball.

Do you see why this is called a "basket star?"  Notice that there are ten arms leaving the central disk.  Although they start branching right away, the initial number is 10, which is a multiple of 5. That is typical of most sea stars.

# Crinoids

**Crinoids** (kry' noydz) can also be mistaken for plants.  These creatures, commonly called **sea lilies** and **feather stars**, have feathery arms on top of cup-shaped bodies.  Unlike the sea star, which generally has its mouth facing down toward the seafloor, the feather star mouth faces up toward the surface of the ocean.  This mouth is surrounded by many feathery tentacles.  Some crinoids have five feathery arms, while others have many more.

As plankton eaters, most crinoids are nocturnal.  However, you may see a feather star during the day with its arms rolled up into a little ball.  Feather star arms are not called rays, but **pinnules** (pin' yoolz).  Each pinnule is coated with a sticky material to catch the plankton that drifts up toward the surface of the sea at night.  A crinoid also has feelers known as **cirri** (sir' eye) attached to the bottom of its body.  These cirri cling to sponges or corals.  Though crinoids look like plants, they do

You can see both the feathery pinnules of this sea lily as well as the cirri that are holding on tightly to the rock in the picture.

not stay in one place like plants do.  They can crawl, roll, walk, and even swim!  However, they only do that if they need to move because of danger or because they are not getting enough food.  If everything is fine, a crinoid can stay in one place, clinging to sponges or corals, waiting for the moment when the nightly plankton swim past.  Unlike a sea star, a crinoid doesn't use its tube feet for walking, only for passing food along its pinnules to its mouth.

*Describe brittle stars and their habits and all that you remember about crinoids.*

# Sea Urchins

Have you ever seen the spiky ball called a **sea urchin**?  This little animal resembles a pincushion with red, purple, green, or other color pins branching out in many different directions.  Unlike the animals we have discussed so far, sea urchins have no arms.  They are, however, completely covered in spines.

The sea urchin's spines protect it from being eaten by most sea creatures.  Who wants a mouth full of needles?  But there are some sea animals that actively hunt for sea urchins.  The sea otter is one example.  The sea otter cracks open the sea urchin with a rock and eats the insides.  You may wonder what is so tasty inside the sea urchin.  Well, often there are millions of

Many sea urchins can fit in the palm of a child's hand.

urchin eggs inside an urchin. Actually, in Japan, people harvest these eggs and sell them for a great deal of money. They are served much like caviar.

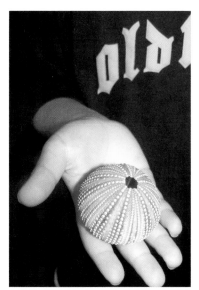

This sea urchin test is what you normally find on the beach.

Under the spines of a sea urchin, you will find a rounded body. When a sea urchin dies, it loses its spines and round, spineless shells commonly wash up on beaches. Although it is okay to call it a shell, it is nothing like the shell of a mollusk. As a result, scientists more properly call it a **test**.

You will often find sea urchins living in close community with other sea urchins. They tend to gather in groups. In tropical waters, the sea urchins have very sharp, needle-like spines that are painful on which to step on. In cooler waters, the spines are shorter and not as sharp.

A sea urchin has tube feet, which it uses to move about slowly. The suction cups grip the rocks so the sea urchin can pull itself along them. Unlike sea stars, these tube feet are found in rows that span from the top to the bottom of the body. As is the case with most echinoderms, its tube feet help the sea urchin find and taste algae or debris, which it then scrapes off the rocks with its five teeth. These five special teeth are arranged in a system that scientists call **Aristotle's lantern**. Some sea urchins can actually use their Aristotle's lantern to bore holes in rocks. They typically bore these holes so that they have a place to hide.

While most sea urchins use their tube feet to move about, there are actually some species that can walk on their spines, using them a lot like stilts!

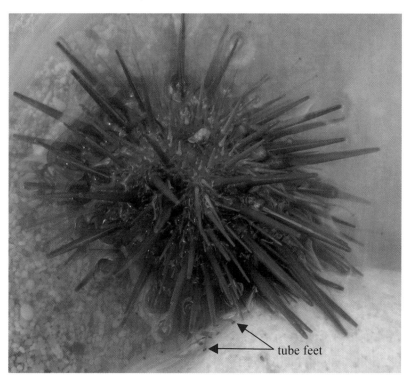

tube feet

The small, stringy strands coming from this sea urchin are its tube feet. Notice how they are gripping the rock with their suction cups.

Most sea urchins are harmless to touch. If you are able to pry a sea urchin loose from a rock, you will enjoy holding it and studying its tube feet, Aristotle's lantern, and its many spines. However, one sea urchin, the **hatpin urchin**, has extremely long spines with serrated tips that, when touched, insert into the skin and snap off. The spine then releases poison. The venom is not fatal, but it is painful, feeling a lot like a bad bee sting. You may want to avoid this particular sea urchin.

God designed some sea urchins to camouflage themselves by holding onto sea grass or ocean debris with their tube feet and carrying it around on their backs. Some even carry live soft corals or anemones for camouflage. They do this to keep from being eaten by other animals.

# Sand Dollars

This living sand dollar looks quite different from the tests of sand dollars that you find on beach.

Another kind of echinoderm is the sand dollar. It is like a flattened sea urchin with tiny spines that remind you of the stubble on the face of an unshaven man. The sand dollars you find on the beach are usually not living sand dollars, but are the shells (once again, more properly called "tests") of dead sand dollars.

If you were to find a living sand dollar, you would notice tiny hairs covering its entire body. These are its spines. They are softer and much shorter than those of the sea urchin.

Like sea stars, sand dollars have tube feet, but they are not used to move around. They are just used to breathe. The sand dollar's tube feet actually stick out the *top* of the sand dollar. If you look at a sand dollar test, you will typically see a pattern that looks like the petals of a flower. That pattern is made up of many tiny holes, and it is through those holes that the sand dollar's tube feet extended when it was alive.

A sand dollar actually uses its tiny spines to move around. It also uses them to dig into the sand. It often does this so that it can bury itself in the sand. Sometimes, a sand dollar will only partially bury itself in the sand, and it ends up poking up from the sand, standing on its side.

The flower shape on this sand dollar test shows you where the tube feet stuck out.

The mouth is at the center of the food grooves in this upside down sand dollar.

On the underside of a sand dollar, you will see a star pattern radiating out from the center where its mouth is located. This star pattern is composed of grooves that are called **food grooves**. Sand dollars filter sand and water, catching plankton and debris on their spines. Then, with the help of tiny hairs called cilia (sil' ee uh), the spines move the food into a food groove, and then the food travels down the food groove to the mouth.

Living sand dollars are typically dark in color, such as brown or purple. Their darker color allows them to stay camouflaged, easily hiding in the sand or mud on the seafloor. This makes living sand dollars hard to find, but someone with a keen eye can usually see a little speck of brown peeking out from under the sand in a shallow pool of water on the beach.

The sand dollars you have seen were probably a pretty white color, because, as I mentioned, the ones we usually find aren't living, but are actually just the tests of dead sand dollars. When a sand dollar dies, it loses its bristles and tube feet. Exposure to the sun then tends to bleach it, producing a nice, white sand dollar test.

How did the sand dollar get its name? Well, many years ago, there was no paper money; all money was in the form of coins. A dollar was usually a very large silver coin. The tests of sand dollars reminded people of these large coins, so people started calling them sand dollars. To this day, some children are told that sand dollars are the money that mermaids use.

*Tell someone all that you have learned about sea urchins and sand dollars.*

# Sea Cucumbers

The last echinoderm I want to discuss doesn't look much like an echinoderm. It looks more like a cucumber that has small bumps or spines on it. I am talking about the **sea cucumber**. No, this is not the fruit of a plant that grows in the sea. It is an animal, and that animal has many of the characteristics of an echinoderm.

You can tell how easy it is to mistake this sea cucumber for the fruit of a plant.

Sea cucumbers are sausage-shaped, benthic animals that can be found in the shallowest of waters as well as the deepest parts of the ocean.  They each have five rows of tube feet that stretch the length of the body, and they use those tube feet to slowly move around the ocean floor.

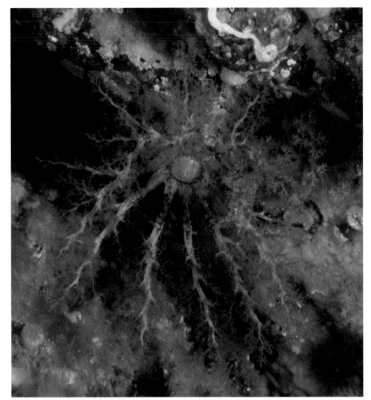

The mouth of this sea cucumber is surrounded by branched tentacles that filter food from the water.

The mouth is on one end of the sea cucumber, and it is usually surrounded by tentacles.  Many sea cucumbers eat sand.  Yes, sand!  The sand enters the mouth, and as it moves through the sea cucumber's system, food particles are absorbed by the body and digested.  Anything that isn't food is eliminated out the other end of the sea cucumber, which is called the **anus**.  Instead of eating sand, other sea cucumbers use their tentacles to filter food out of the water.  Then, they move the food to the mouth, where it can be eaten.  If a sea cucumber filters water in this way, its tentacles are usually longer and more branched than the tentacles of a sea cucumber that eats sand.

One of the most fascinating things that sea cucumbers do is spill their guts.  No, they don't tell you about all their problems.  When a sea cucumber spills its guts, it literally spills its guts out.  When threatened by a possible predator, the sea cucumber tosses out long sticky threads that are actually made up of its own internal organs.  These sticky threads can glue a predator's gills or throat shut, which will eventually kill the animal.  It certainly gives the predator something else to worry about instead of eating the sea cucumber.  That's a nice form of defense, isn't it?  Of course, these body parts are soon regenerated, so there is no long-term damage to the sea cucumber.

Sea cucumbers come in a variety of colors, and can be as small as your fingernail or several feet long.  You may find one if you frequent tide pools or reefs, but don't be surprised if you mistake it for a fruit!

# What Do You Remember?

What is the name of the phylum that includes sea stars, sand dollars, sea urchins, and sea cucumbers?  What does "phylum" mean?  What is special about these animals' feet?  Explain how sea stars eat.  What is a sea star's favorite food?  What did clam fishermen once do to keep sea stars from eating the clams?  Why did this not work?  Why are brittle stars considered brittle?  How do they move across the ocean floor?  How is this different from sea stars?  What are sea urchins' teeth called?  What animal really likes to eat sea urchins?  Explain how a sand dollar eats.  How does a sea cucumber defend itself?  In what two ways do sea cucumbers eat?

# Notebook Activities

Make a notebook page labeled "Echinoderms."  List the things that most echinoderms have in common.  Then, illustrate each of the following animals and write down what you have learned about them: sea stars, brittle stars, crinoids, sea urchins, sand dollars, and sea cucumbers.  The *Zoology 2 Notebooking Journal* includes pages to complete these assignments.

**Older students**: Copy and memorize the poem below about the legend of the sand dollar. A page for you to complete this activity is provided in the *Zoology 2 Notebooking Journal.*

## Legend of the Sand Dollar[*]

There's a pretty little legend
That I would like to tell
Of the birth and death of Jesus
Found in this lowly shell.
If you examine closely,
You'll see that you find here
Four nail holes and a fifth one
Made by a Roman's spear.
On one side the Easter lily,
Its center is the star,
That appeared unto the shepherds
And led them from afar.
The Christmas poinsettia,
Etched on the other side
Reminds us of His birthday,
Our happy Christmastide.
Now break the center open,
And here you will release
The five white doves[**] awaiting
To spread good will and peace.
This simple little symbol,
Christ left for you and me
To help us spread His Gospel
Through all eternity.

[*]We do not know who originally wrote this poem.
[**]The "five white doves" mentioned in the poem are the five teeth that make up the sand dollar's Aristotle's lantern.

# Ocean Box

Add echinoderms to your ocean box. To make a sea urchin, you can cut toothpicks in half and insert them into a ball of clay or a Styrofoam ball. Sea stars can be made from clay. If you have small craft feathers, or a feather pillow that can afford to lose a few feathers, you can use them to make a feather star.

# Project
## *Make a Salty Brittle Star*

Today we are going to have a little fun with salt water. We are going to create salty brittle stars by crystallizing salt.

**You will need:**
- Five 6-inch segments of cotton string
- Three cups of water
- A lot of salt
- Food coloring (any color)
- A wide-mouthed glass jar (A glass measuring cup will do.)
- A pan for boiling water
- A wooden spoon for stirring

1. Boil the water in a pan.
2. Once it starts boiling vigorously, remove it from the stove and add a few drops of food coloring.
3. Pour in some salt and stir to dissolve.
4. Continue adding salt and stirring until no more salt will dissolve into the water. This will produce a **saturated** saltwater solution. Be sure that no more salt can dissolve in the water. The project won't work unless the solution is saturated.
5. Pour the solution into the wide-mouthed jar. If you have extra solution, put it in a glass and save it for later.
6. Tie your strings together, making a knot where they all come together. Place the knot of your string into the container. Position the strands of the string so that the ends hang over and around the rim of your container, as shown in the picture.
7. Place the container in a safe place where it will not be disturbed for a couple of days.
8. Watch what happens. In a couple of days, you should see salt crystals begin to form around the

strings. If the solution begins to evaporate below the knot, add the extra solution you saved. In the end, you should have a salty brittle star!

# Lesson 12
# Cnidarians

Jellyfish are cnidarians that have the medusa form as adults.

While walking through a field, have you ever brushed against a plant that had sticky spines on it? The spines lodged in your skin, burning and itching for hours. One such type of plant is called the **nettle**. It's a terrible plant to encounter. Though it just sits there appearing to be nothing harmful, when you rub against it, you know it's something quite harmful.

Well, like some plants, there are also some nettle-like animals that are best to avoid. Though they either drift or sit in place seeming like nothing harmful, they possess a powerful sting that can cause people pain and can kill the creatures that the animals want to consume. These animals are called **cnidarians** (nih dahr' ee uhnz), because they belong in phylum **Cnidaria** (nih dahr' ee uh). The name comes from the Greek word "*knide*," which means "nettle." And just like the plant called nettle, cnidarians pack a powerful sting. Jellyfish, sea anemones, and corals are the cnidarians we will study in this lesson.

# Polyp vs. Medusa

Boneless, brainless, eyeless, headless, and footless, cnidarians are mostly composed of a mouth and tentacles. Some cnidarians have tentacles that hang below them, like strings dangling from a bell. Other cnidarians have tentacles that wave above them, like little arms waving in the air. Those with waving tentacles are called **polyps**, and those with dangling tentacles are called **medusae** (meh doo' say — the plural of **medusa**).

Polyps have stocky, tube-shaped bodies with tentacles that look like lots of little feet. In fact, the word "polyp" comes from a Greek word that means "many feet." Notice the hole in the center of the sea anemone on the right. That's its mouth.

A sea anemone is a cnidarian that takes the polyp form. Notice the mouth at the center of the tentacles.

Besides the direction in which the tentacles hang, medusae and polyps are different when you compare their behavior. When they are medusae, cnidarians tend to float in the sea and are considered plankton. Polyps, on the other hand, generally stay attached to one place. They can move if they need to, but as long as food is plentiful, they say put. It should be noted that, some cnidarians are *both* polyps and medusae. They start off their life in one form, and then they switch to the other form as they mature. Many jellyfish, for example, begin their lives as polyps but then mature into the medusae that we generally think of when we think of jellyfish.

All these dangling or waving tentacles are equipped with cells that kill or paralyze the cnidarian's prey. Their stings can be quite a bit more painful than those from a nettle bush, and a few species of jellyfish are even deadly to people.

# Nematocysts

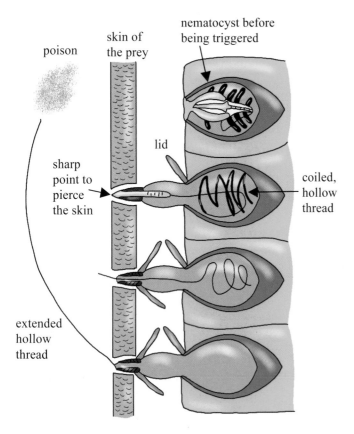

poison

skin of the prey

nematocyst before being triggered

lid

sharp point to pierce the skin

coiled, hollow thread

extended hollow thread

Let's take a closer look at a cnidarian's intriguing stinging cells, which are called **nematocysts** (nih mat' uh sists). The nematocysts typically cover each tentacle; there could be thousands or even millions of these little cells on one tentacle, depending on how big the tentacle is. Each nematocyst is like a small jar containing an arrow with its own poison delivery system.

This poison arrow jar has a little lid that is closed until something triggers it. In many cnidarians, nematocysts are triggered when the tentacle is touched. Other cnidarians have a sophisticated system that triggers the nematocysts only when a particular chemical is detected. Once it is triggered, the lid to the nematocyst's "jar" (called the **capsule**) springs open, and the arrow within pops out.

The arrow's job is to pierce the skin so that the nematocyst can extend a long, hollow thread into the animal. This thread is coiled up in the nematocyst's capsule, but once the sharp tip has pierced the skin, the thread shoots into the prey's body. When the thread enters the body and is extended, a powerful poison is released from the end. All of these events happen *incredibly* quickly. Believe it or not, once the nematocyst is triggered, it can take as little as *one thousandth of a second* for the capsule to open, the sharp point to pierce the skin, the thread to be inserted, and the poison to be released!

All the nematocysts that are triggered are released at once. As a result, there could be thousands of nematocysts popping open, flinging their poison threads into the victim. This usually kills the prey or at least puts it into shock, which basically means the prey is helpless.

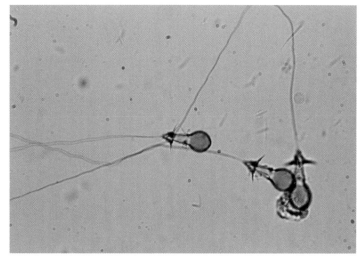

This photo of discharged nematocysts was taken using a microscope. The nematocysts are magnified 450 times their normal size.

After the nematocysts have done their job, the tentacles pull the prey to the cnidarian's mouth. The prey is then forced into the mouth — either the whole thing if it's a small animal, or bit by bit if it's larger, as the tentacles break it down into smaller pieces. Once the poison arrow is spent, a nematocyst will not produce any more arrows. Instead, it breaks down and is absorbed into the cnidarian's body. Then, a new nematocyst grows in its place.

Most sea anemones and coral polyps have short-stringed nematocysts that can't penetrate all the way through a person's skin. Nevertheless, they easily penetrate smaller prey like plankton and small fish. Jellyfish, on the other hand, have nematocysts that can penetrate human skin. As a result, while a person will not suffer from the sting of most corals and sea anemones, the sting of a jellyfish can be painful. Some can even be deadly. If you happen to be stung by a jellyfish, try to get some vinegar on the stung area immediately. This neutralizes the nematocysts' poison, reducing the pain that they cause.

# Creation Confirmation

Some scientists think cnidarians are "simple" animals. These scientists consider cnidarians to be early animals, uncomplicated in their design because they were produced by natural processes very early in earth's history. However, their nematocysts are some of the most complex animal structures in all of creation. They fire quickly and accurately with devastating consequences to the prey. How in the world can scientists think that the animals that *build them* are simple? Remember, once a nematocyst is used, it decays away, but the cnidarian that used it simply builds another one in its place. In other words, these supposedly "simple" animals not only have fully functional weapons they can use to attack prey, but when those weapons are spent, they just build more. Can you think of any weapon made by our military that can continually regenerate ammunition when needed? Even the best of human technology pales in comparison to what these "simple" animals can make. This shows us that these animals are *not* simple. They are elegant creations, designed by God.

*Can you explain in your own words how nematocysts work?*
*What is the difference between a polyp and a medusa?*

# Jellyfish

This white spotted jellyfish (*Phyllorhiza punctata*) is commonly found around Australia, especially during the summer months.

**Jellyfish** are cnidarians that usually prefer to hang out in warm water, the same as people do. They move with the currents, because they are zooplankton. Some jellyfish can't swim at all. They simply float in the water, and get carried wherever the currents go. Other jellyfish can swim, but they cannot swim strongly enough to overcome the currents. These jellyfish have *some* say in where they are going, but in the end, the currents really determine where they will end up. Some currents take jellyfish out to sea, while others bring them right up onto the beach where they can get caught in waves and are thrown onto the sand. They really can't choose where they will end up. So, if a jellyfish stings you, it wasn't chasing you down. Either you accidentally bumped into it, or it accidentally bumped into you. Once that happened, its long tentacles only did what came naturally.

Despite the fact that jellyfish don't have brains to process visual information, they do have eyes that can detect light from dark. This allows jellyfish to discern up from down, which is very important for those that can swim. A jellyfish that can swim pumps its medusa open and closed, which pushes water away from the jellyfish. As a result, the jellyfish moves in the opposite direction. Generally, a jellyfish does this to escape predators by rising to the surface at night and sinking to the deep ocean waters during the day. Of course, near the ocean floor, they have to watch out for sea stars, which love to eat jellyfish. Their main predators, however, are squids and sea turtles. Do you remember which sea turtles hunt jellyfish? That's right. Leatherback sea turtles *love* to feast on jellyfish.

It is amazing that the huge leatherback turtle survives on a diet of jellyfish. After all, jellyfish are more than 95% water! If you took a 400-pound jellyfish (yes, some jellyfish do grow to be that heavy) and removed all the water from it, it would weigh fewer than 20 pounds! The jellyfish is basically two layers of skin. The outer layer is called the **epidermis** (ep' ih dur' mis), which is the same name scientists use when they talk about your skin. The inner layer is called the **gastrodermis** (gas' truh dur' mis). In between these two layers is a jellylike substance called **mesoglea** (mez' uh glee' uh).

If you ever find a jellyfish washed up on shore, look for a fish inside its transparent body. If the fish was eaten a while ago, it will appear white because most of the nutrients have been absorbed, but you can still discern its shape through the transparent skin of the jelly. The favorite food of most jellyfishes is fish.

Jellyfish can be as small as the tip of your finger, with teeny-tiny tentacles, or they can be as large as a huge patio umbrella, with tentacles that dangle hundreds of feet below! Do you remember that the lion's mane jellyfish rivals the blue whale as the longest animal in creation? It has tentacles that can be over 100 feet long! However, you need not worry; you aren't likely to encounter this enormous creature unless you spend time in the deepest parts of the ocean off the coast of Japan. Thankfully, most jellyfish have medusae that are about the size of your face, and their stings are uncomfortable but cannot cause serious harm to people.

Jellyfish tend to congregate in swarms. These swarms are called **smacks**. If a smack gets very large, it can be miles long and contain thousands and thousands of these creatures. When a smack gets that large, it is known as a **jellyfish bloom**. No one is sure what causes a jellyfish bloom. Most jellyfish blooms are found far out at sea. However, lately, Japanese waters are swarming with giant jellyfish. The Japanese consider it an invasion. When fishermen try to haul in their catches of fish, the nets break under the weight of the many large jellyfish that are caught as well. This affects the entire region because it makes fish, the main food of the Japanese, harder to catch. In fact, the countries of Japan, China, and South Korea are working together to figure out what to do about these creatures. No one is sure what the sudden surge of giant jellyfish is all about.

Jellyfish tend to swarm together in large groups called "smacks."

# Making More Jellies

As is the case with many creatures in the sea, young jellyfish typically look nothing like their parents. After two jellyfish mate, the eggs that are produced usually stay attached to the mother until they hatch into larvae. The larval form of a jellyfish is called a **planula** (plan' yuh luh), and it is a tiny creature covered with even tinier hairs. A planula leaves its mother and floats with the currents, eventually settling down to the seafloor where it attaches to something hard. There, it becomes a polyp, looking something like a sea anemone. After a while, the polyp grows buds that develop into tiny jellyfish, which detach from the polyp and drift away. At that point, it looks like its parents.

# Creation Confirmation

In 2003, scientists found a huge group of large jellyfish fossilized in a Wisconsin sandstone (stone formed from sand) quarry. This was an amazing discovery because jellyfish that get stranded on the beach quickly decay due to their thin, water-filled bodies. In fact, as soon as the sun hits a jellyfish on the beach, it begins to dry out and quickly disintegrates to nothing but a thin skin, which then deteriorates rapidly. In order for these fossils to have been made, then, the jellyfish must have been buried by sand immediately, before they had time to decay. Thus, whatever event caused these jellyfish to become fossils had to be quick and had to involve a *lot* of sand.

During the worldwide Flood spoken of in the Bible, large masses of sand would have been carried around by the churning flood waters. Those masses of sand would quickly cover whatever got in their way, causing a perfect situation for fossilization of even the most fragile of animals. This discovery of fossilized jellyfish is evidence for that Flood. Those jellyfish were probably buried by large amounts of sand very quickly. That is the best way to explain how such easy-to-decay animals were fossilized.

# Floating Boxes

The box jellyfish is potentially deadly to people, but it is not a true jellyfish.

Although most jellyfish cannot cause permanent harm to people, a sting from certain jellyfish can prove fatal. Many of the "jellyfish" having a potentially fatal sting, however, are not actually jellyfish at all. You see, there are some animals that look like jellyfish but are not *true* jellyfish. A true jellyfish has an umbrella-shaped medusa. If a creature looks like a jellyfish but does not have an umbrella-shaped medusa, it is probably not a true jellyfish. Instead, it is a jellylike cnidarian. Consider, for example, the box jellyfish in this picture. The poison contained in its nematocysts is strong enough to kill a person, but you can tell by the boxy shape of its medusa that it is not a true jellyfish. In Australia, beaches can be closed down if a smack of box jellyfish is seen nearby.

# Floating Friends

Some jellylike cnidarians don't like to live alone, but forming swarms isn't enough for them. Instead, they like to live in a colony, with many individuals helping one another to survive. Typically, a single individual will reproduce until there are several clones of itself. Because the clones are all attached to one another, they look like one single animal, but there are really many, sometimes hundreds, of different animals living as one.

One example of such a colony is the **Portuguese man-o-war**.  Although many people mistake it for a jellyfish, it is not.  It is actually hundreds of jellylike creatures (not true jellyfish) all linked together.  The Portuguese man-o-war is found in warm waters off the coasts of Africa, North America, Europe, and Australia.

This colony of animals has tentacles that are able to hang up to 100 feet below the surface.  They pack a powerful sting, too.  Their sting is not as bad as the box jelly, but almost.  If a person gets stung by too many of a man-o-war's tentacles, he or she might go into shock.  Also, vinegar can make this colony's sting worse, so you can't treat it like a jellyfish sting.

Each creature within the man-o-war colony has its own job to do.  One creature's job is to be the sail.  The sail polyp fills its body, which can be 6 to 12 inches long, with gas and keeps the entire colony floating on the

This Portuguese man-o-war is actually a colony composed of many animals.

ocean, being carried here and there by the wind.  This polyp gives the man-o-war its distinct blue or purple bottle-shaped look.  Other polyps have jobs like stinging, eating, digesting, sharing the nourishment of a meal with the entire colony, or reproducing to make more polyps.  They all work together to keep the entire colony alive.

*Explain in your own words all that you have learned about jellyfish and jellylike cnidarians.*

# Sea Anemones

**Sea anemones** (uh nem' uh neez) are the flowers of the sea.  In fact, they are in a class of animals called **anthozoa** (an' thu zoh' uh), a word which means "flower-animal."  Like the jellyfish, sea anemones have tentacles that contain nematocysts.  Anemones use their tentacles to kill or paralyze small fish or other unsuspecting animals that happen to pass too closely.

Sea anemones differ from jellyfish in a few ways.  First, the nematocysts of most anemones can't penetrate into human skin, so you can't feel it if they try to sting you.  As a result, you can pick up and study most anemones and never feel a sting.  Second, anemones spend their entire lives as polyps.  Remember, jellyfish start out life as a planula that develops into a polyp, but that polyp eventually produces medusae.  Sea anemones start out as polyps and stay polyps all of their lives.

Sea anemones are often brightly colored.

Yet another way anemones differ from jellyfish is by their brightly colored bodies. Like flowers, they come in an array of lovely colors and patterns: purple, red, yellow, orange, pink, blue, and even spotted or striped!

Unlike the planktonic life of most jellies, anemones are benthic. Yet, these little flowers of the sea can actually move. Even though an anemone usually attaches itself to a hard surface and stays there, it can move around by scuffling along on the base of its stalk. Some can even swim. Others can get where they want to go by somersaulting along the seafloor.

Sea anemones prefer a nice, high spot on rocks and corals to catch the most plankton or other animals that happen to float or swim past. Have you ever played "king of the mountain?" Children have played this game for years. One child is on top of a hill and remains the king until someone else nudges him off. If you don't know how to play, just spend some time with sea anemones. They'll teach you. Sea anemones war against each other for the best spot on the tallest rock. You'll see them pushing and pushing against other anemones to topple them off their perch. As soon as it has a victory, the winner takes its place as king of the mountain. But not for long, as others will try to claim the throne.

Anemones react quickly when touched. In fact, when lightly touched, the entire anemone will contract, sometimes closing up if it feels threatened. The ability to close up is very helpful, as sometimes anemones find themselves too close to the shore when the tide goes out. Thankfully, if the anemone is left with no water covering, it turns its entire body inside out, closing its tentacles inside itself. In this position, it can protect itself from drying out by preserving water inside its body. This is a defense system that protects the anemone until the tide comes back in. Jellyfish don't have this defense system, which is why they will die before the tide returns if they are stranded on the shore.

Though they can't see, anemones can sense movement and know when something good to eat brushes past their tentacles. If that happens, a sea anemone will shoot its nematocysts into the prey. Once the prey is paralyzed by the nematocysts, the anemone will use its sticky tentacles to grab hold of the animal and drag it into its mouth.

If you were to spread open the tentacles of a sea anemone to see into its middle, you would find a hole right in the very center. That hole is its mouth. Stunned prey will be pushed down into the anemone's mouth to be digested. Any parts the anemone can't digest will be thrown out the same mouth hole. If you place your finger in an anemone's mouth, it will close its stomach around your finger. Don't worry, you're safe. It won't eat your finger. You can simply pull your finger right back out.

Most anemones are about the size of your hand, but they can range in size from as small as a nickel to as big as 2 feet in diameter. That's as big around as a couch cushion!

This sea anemone's mouth is clearly visible at its center.

## Anemone Associates

The clownfish is safe inside the tentacles of a sea anemone.

Even though sea anemones taste bad to many animals and produce dangerous stings to those who come near, they have some friends in the sea. Certain animals produce a special coating of slime that keeps them from getting stung by the anemone. The anemone's tentacles become the animal's safe home; they form a wall of protection around the creature, where no predator would dare come.

The anemone and the animal living within it form a symbiotic relationship with one another, each giving something to the other. The anemone gives the animal protection, while the animal gives the anemone something in exchange. Are you wondering what the animal gives the anemone in exchange? Well, different animals give the anemone different things. The clownfish, also called the anemonefish, chases away butterfly fish that feed on sea anemones. The clownfish also eats parasites off the anemone, keeping it clean. In addition, it attracts predator fish to the sea anemone so that the anemone can catch and eat them. Certain shrimp that live within sea anemones, called cleaner shrimp, eat bacteria and algae that try to grow on the body and tentacles of the anemone. There is even a crab called the **boxer crab** that attaches an anemone to each cheliped and walks around using the anemones to sting small animals in order to eat them. The anemones then share in the feast.

# Adding Anemones

Sea anemones reproduce in two ways. The first way is the normal one: males and females mate to produce larvae. The larva floats around until it settles onto a hard surface, and then it starts to develop into a polyp. The second way a sea anemone reproduces is by **budding**. When a sea anemone reproduces in this way, a small "bump" grows on the anemone's body. That bump grows bigger and then pinches off from the body, settling onto the seafloor and growing into a separate polyp. This kind of reproduction produces an exact copy, or clone, of the original sea anemone. Do you remember another animal that can make a clone of itself? The sea star can make a clone of itself.

Whether produced by normal reproduction or by budding, a new sea anemone can live a long time. If you find an anemone on the beach, it might be older than you or even your mother. Why, it might even be older than your grandmother! Anemones have been known to live up to eighty years!

*Explain what you have learned so far about anemones.*

# Coral

Do you know what **coral** is? Look at the picture on the right. It's a picture of coral. You may be wondering why on earth I'm including it in a book about sea animals. Would it surprise you to learn that corals are actually animals? Well, what you actually see in this picture are the *remains* of hundreds or thousands of once-living corals. The coral animals that made these remains are much like tiny sea anemones.

This is a set of remains from once-living coral.

Well, not all corals are tiny. Some are as big as your dad's foot, but most are rather small. Like the anemone, a coral is a polyp that has a tubelike body with a hole in the middle that forms the mouth, and that mouth is surrounded by stinging tentacles. Because they resemble sea anemones, corals are in the same class as sea anemones, class Anthozoa. Corals, then, are also "flowers of the sea."

The most important difference between sea anemones and coral polyps is that many coral polyps build protective walls around themselves. In other words, they build their own houses. Those

Coral reefs can be incredibly colorful.

"houses" are left behind when the corals die, and that's what you typically find on the seashore. Although some people will call these remains coral shells, they are not shells. It is more proper to call them **coral skeletons**. Each different kind of coral colony builds a different kind of skeleton, resulting in colonies that can be shaped like a brain, a mushroom, a cabbage, or many other things. With all these corals gathered together building skeletons around themselves, large structures are formed. We call these coral formations **coral reefs**.

Corals that build hard skeletons around themselves are called **stony corals**, but it is important to realize that not all corals are stony corals. Some are **soft corals**, which are soft and can be easily mistaken for plants. We will study those in a moment. The corals, whether stony or soft, come in an amazing variety of beautiful colors. They truly make the tropical waters where they grow a delightful, rainbow-colored underwater scene.

# Stony Corals

Stony corals are typically called **reef-building corals**, because they are largely responsible for making coral reefs. How do these reefs begin? Well, when a reef-building coral is first hatched from an egg, the larva swims around until it finds a place to begin growing. It usually attaches to some rock or another piece of coral. Immediately, the polyp begins to make more polyps by budding, just like a sea anemone. It does this again and again until there are many, many coral polyps.

Each coral polyp in the colony is connected to every other coral polyp through living tissue that stretches from polyp to

This close-up of a living hard coral colony shows you the individual polyps and their tentacles.

polyp. So when one polyp catches food, it can send the food particles around so that they feed all the nearby corals.  Those corals can then share with corals near them, and on and on, so that food is passed throughout the colony.  In this way, even if one coral polyp catches no food for the day, it will eat when another polyp does.  This makes the polyps in a colony dependent on one another for survival.

Each coral polyp makes its hard, outer skeleton by converting calcium in the water into calcium carbonate, the main component of chalk.  The cuplike skeleton grows into the different shapes of the coral colony.  To capture prey, the individual polyps stick tentacles out of holes at the top of the cup.  When danger approaches, a polyp swiftly pulls back into its cup.  Most corals stay in their cups all day and only emerge at night to eat.  Can you guess why they feed at night?  As I told you in previous lessons, that is when millions of zooplankton make their way up to the surface of the ocean.  Those zooplankton make an easy meal for the corals.

# Assisting Algae

Another interesting thing coral polyps do is capture tiny phytoplankton called **zooxanthellae** (zoh zan thel' lee).  But interestingly, instead of being eaten, the zooxanthellae are kept inside the coral polyp's body so that they can make nutrients for the polyp.  The polyp doesn't *have* to have these nutrients to survive, but they make the polyp healthier. They act like fertilizer that you might put in your garden.  The plants don't need the fertilizer, but it does make them healthier.  Another thing the zooxanthellae do for the coral is increase the speed with which it can build its outer skeleton.  Without zooxanthellae, it would take the coral polyp a *lot* longer to make its protective home.  Of course, the zooxanthellae benefit from this symbiotic relationship as well, because they are protected by the coral's outer skeleton.

Notice the parts of the coral that are bleached.  If the white polyps can regain zooxanthellae, they will return to the same color as the rest of the coral in the picture.

Because zooxanthellae are phytoplankton, they use light from the sun to make their own food.  As a result, corals usually live in water that gets a lot of sunlight.  You rarely find corals in the deep, dark depths of the ocean, where no light penetrates.  You also rarely find corals in water that is murky and dark.  Rather, you usually find corals in clear water.  In addition, coral polyps need warm water to produce their outer skeletons, so the water not only has to be clear, it needs to be warm as well.

Another thing the zooxanthellae do for coral is provide the coral with its color. If the zooxanthellae in the coral die, the coral loses its coloring.  Scientists call this **bleached** coral.

The bleached coral can live without the zooxanthellae, but it will not be nearly as healthy as it could be.  If the coral can manage to capture more zooxanthellae, it will return to its normal color.

# Coral Reefs

Eventually a coral polyp will die.  However, though the animal inside dies, the outer skeleton that protected it remains.  In addition, as a polyp dies, a new polyp from the same colony will be produced by budding, and it will put its new home on top of the dead coral!  As time goes on, this process continues, forming a large structure called a coral reef.

Coral reefs are teeming with life!

Some coral reefs have taken hundreds or even thousands of years to form.  The largest coral reef, which is near Australia, is called the **Great Barrier Reef**.  It is more than 12 hundred miles long and is several hundred feet thick in some places.  Creation scientists estimate that it would have taken about 3,500 years for this reef to grow to the size it is today!  Now remember, even though the Great Barrier Reef is very thick, only the topmost layer contains living coral.  The rest of the reef is made up of the dead coral remains on which the living corals built their homes.

Coral reefs are an important habitat for sea life.  Millions of animals depend on the reef for food and protection every day of their lives.  Coral reefs are actually like the rainforests of the sea, housing some of the most colorful, exotic animals in the ocean.  This makes reefs a favorite spot for divers to visit.

Coral reefs support so much sea life because they provide a lot of food for some animals.  Crown of thorns sea stars, nudibranchs, parrotfish, butterfly fish, and other animals eat coral.  In addition, coral reefs provide homes and safe hideouts for other animals.  Coral reefs are dotted with many crevices and holes.  Fish, crabs, shrimp, sea horses, and other animals make their homes in those crevices and holes, trying to hide from predators.  Of course, with all of these animals eating and hiding in coral reefs, large predator fish are attracted to coral reefs, hoping to eat the animals that are eating the coral or find the animals that are hiding in the coral.

# Reefs at Risk

Because corals depend on warm, sunlit water to survive, coral reefs are often found in shallow waters that are close to the beach. Unfortunately, being so close to land makes them vulnerable to poisons and sediment that can come from boats and the land nearby. Sediments make the water more cloudy, which reduces the sunlight that the zooxanthellae can use. Poisons, of course, kill the corals, and they can also bleach the corals. In addition, ships run into the coral, damaging the coral and the boats. All of these dangers must be limited if reefs are to survive. As a result, many countries are trying to limit the kinds of construction that can occur near beaches that have nearby coral reefs (which will reduce the sediments and poisons to which the reefs are exposed) as well as the kinds of boat traffic that can occur around reefs.

# Soft Corals

Soft corals get their names from the fact that they don't form hard skeletons around themselves. However, most do have tiny limestone spines imbedded in their skin. This helps support the animal as it grows and provides some protection against predators. Because they do not have to build hard homes around themselves, soft corals grow faster than stony corals. However, because they don't have those hard homes for protection, they are more likely to be eaten by predators or damaged by the motion of the sea.

The flecks you see on this soft coral are the limestone spines it makes. The flower-like structures are the polyps themselves.

Many soft corals look like underwater trees that sway in the ocean currents. If you didn't know better, you might think they were plants, not animals! Of course, they really are animals, because they cannot make their own food. Instead, just like stony corals, they capture food with their tentacles.

# Non-nettle Jellies

Ranging from the size of a pea to the size of your entire arm, the **comb jelly** looks a lot like a jellyfish, but it is not a cnidarian. Do you remember what the word "cnidarian" means? That's right; it means "nettle animal." Comb jellies aren't cnidarians because they don't have nematocysts. Thus, they don't have a sting. Instead, this little creature has a sticky substance on its tentacles to capture prey. Small creatures get stuck in these sticky tentacles and end up as lunch for the comb jelly when the jelly sweeps its catch up into its mouth. Unlike most jellyfish, the comb jelly can actually retract its tentacles when it is not feeding.

Even though comb jellies are not in class Cnidaria, I included them here because they are so similar to jellyfish. Though numerous in the sea, they are hard to see because they are so transparent. Unlike most other creatures, these little rascals live both near the shore and also in the deepest, darkest depths of the sea. Down there, they are some of the most plentiful animals around.

Why are these animals called comb jellies? Because of the combs of course! In science, the common name usually has something to do with what the animal looks like. All around the body of a comb jelly are eight columns of tiny hairs in such neat rows that they look like combs in the animal's body. The hairs on each comb wave back and forth so that the comb jelly can control which way it is pointing as it floats. Even though it can control how it points in the water, the comb jelly is considered to be a part of the plankton community, because it cannot swim well enough to fight the ocean's currents.

A few kinds of comb jellies have bioluminescence. Do you remember what that is? It is the ability to make light, like a firefly. At night, these comb jellies are visible on the surface of the water as millions of tiny lights. Small unsuspecting creatures may swim toward the light and end up inside the comb jelly's stomach.

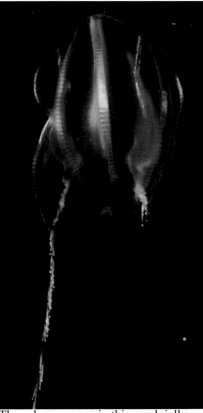

The colors you see in this comb jelly are due to how light reflects off the combs. Notice how the left tentacle is extended for feeding, but the right tentacle is retracted.

# What Do You Remember?

What does "Cnidaria" mean? Where is the mouth on a cnidarian? Explain how nematyocysts work. What is the difference between a polyp and a medusa? Why are jellyfish considered plankton? What is special about the box jelly? What is special about the man-o-war? How do sea anemones and corals differ from jellyfish? How do corals differ from sea anemones? Where do corals grow and why? Explain how corals are dependent on algae. What is the difference between stony coral and soft coral?

# Notebook Activities

Write down what you learned about jellyfish, sea anemones, and corals. Make an illustration of each animal for your notebook. The *Zoology 2 Notebooking Journal* has pages for this activity.

**Older Students:** Research where the Great Barrier Reef is located. Draw it on a map and place this in your notebook. A page for this assignment is included in the *Zoology 2 Notebooking Journal*.

# Ocean Box

You will want to add at least three of the animals we discussed (a jellyfish, a sea anemone, and a coral colony) to your ocean box. To make a jellyfish, you can use clear plastic wrap. Form an umbrella shape for the medusa, and then twist several strands to make the tentacles. Attach the tentacles to the medusa with clear tape. If you choose to do a man-o-war, you can blow up a small balloon and attach twisted plastic wrap to the bottom of it. To make a sea anemone, you could use a piece of clay with small pipe cleaners for the tentacles.

# Experiment

Jellyfish and all plankton are completely dependent on currents in order to move through the water. Surface currents, called gyres, move animals around near the surface of the ocean. Do you remember where many zooplankton spend the daylight hours? They spend it in the deep ocean, hiding from predators. Well, you may remember that there are currents down in the deep part of the ocean, too. Those currents are pushing the plankton that live down there as well. In the first lesson, we made gyres that moved the water around on the surface. Now let's create a deep sea current so we can see what carries these deep-water plankton around.

**You will need:**
♦ A "Scientific Speculation Sheet"
♦ A glass baking dish
♦ Hot water
♦ A fruit ice pop (For best results, use one made from real fruit, like an Edy's frozen fruit bar.)

1. Fill the glass baking dish with very, very hot water.
2. While the water settles down, make a hypothesis about what will happen when you put the ice pop in the water on one side of the baking dish. Record it on your "Scientific Speculation Sheet."
3. Gently place the ice pop in the water on one side of the baking dish.
4. Record what happens.

What happened in the experiment? The ice pop was very cold, but the water was very hot. This caused the ice pop to melt. As it melted, you should have seen the liquid from the ice pop begin to spread throughout the baking pan. Did you notice how it flowed, however? It flowed along the bottom of the pan. The temperature difference between the ice pop and the hot water caused the current that made the liquid flow, and because the liquid from the ice pop was cold and had sugar in it, it was denser than the water, so it stayed near the bottom of the baking dish. Much like what happened in your experiment, the deep water currents that push plankton around are caused by temperature differences between the cold water that is deep in the ocean and the warmer water that is near the surface.

# Lesson 13
# Other Interesting Aquatic Animals

Well, your study of aquatic creatures is nearly complete. You have gained a world of knowledge, but there are still more critters to be discussed. Some scientists call the animals we will study in this section "simple animals," because they do not have some of the body parts that we normally think animals should have, such as eyes and ears. Others, like the tubeworms pictured on the right, really look like plants, never moving anywhere. Still others are so tiny that you need a microscope to see them.

You may be wondering how a scientist knows if a creature is an animal, a plant, or something else (like a bacterium). Well, when examining an animal under a microscope, scientists can easily tell if it's a plant or an animal by examining the **cell** or cells that make up the creature. If you don't know what a cell is, it is the basic unit of a living creature. Cells

Although you might think these are underwater plants, they are animals called tubeworms.

are so tiny that they can only be seen under a microscope. Anything that is living must have at least one cell. Of course, if it has only one cell, you can't see it without the help of a microscope. The animals you have been studying in this book are made up of many, many cells, which is why you don't need a microscope to see them.

Animals and plants are made up of many cells. As a result, we say they are **multicellular**. "Multi" means "many," so "multicellular" means "many cells." In fact, animals and plants are made up of many *different kinds* of cells. An animal's skin is made up of one kind of cell, for example, while its heart is made up of other kinds of cells. There are other creatures, like bacteria, that are **unicellular**, which means they are made of only one cell. These creatures are so small that you cannot see them without the help of a microscope, but they are, indeed, living things.

So, to be an animal or plant, a creature must be multicellular. If a multicellular creature consumes food and has the ability to move on its own, it is an animal. You see, animals are **consumers**; they must consume food to live. Plants don't move around and don't consume food; they make food. Because plants make food for themselves, we say that they are **producers**, because they produce their own food.

Have you ever seen a plant walk around?  No, because a plant can't move around.  Do you know what a mobile home is?  It's a house that you can move around to different locations.  The word "mobile" means that it can be moved from place to place.  Animals are mobile, but plants are not.

So how would you describe an animal?  See if you can do it.  As you just learned, scientists say that an animal is a multicellular, mobile consumer.  That's the definition of an animal.  Sounds kind of scientific doesn't it?  Can you tell me what each word in the definition of an animal means?  Wow — five minutes ago, do you think you would have known what those words meant?  Well, now you do. You have learned a lot already!

# Sponges

These sponges are actually animals!

If you were scuba diving in the ocean and came upon the things pictured on the left, would you be surprised to learn that you were looking at animals?  They are called **sponges**, and there are thousands of different species of them in the ocean. You probably have used sponges before.  Most likely, you have used them to wash dishes or even wash yourself in the bathtub.  Although the sponges you have used were probably synthetic (which means that people made them), they are actually modeled after these soft, squishy animals.

Actually, not all sponges are soft and squishy like the ones we use.  Some are hard and prickly. Sponges come in many shapes and sizes.  Some are as tiny as your finger, and some are as large as a shower stall.  Some look like tubes, while others look like tree branches, and they all look more like plants than animals.  A sponge has no eyes, no ears, no brain, no nerves, no heart, and no blood.  As an adult, it spends its entire life attached to an underwater rock, never moving from that spot.  When something is stuck to one spot, we say it is **sessile** (ses' eyel).  Well, sponges are sessile.  But even so, sponges are not plants, because they are consumers.  Still, they do seem rather "un-animal-like," don't they?  In fact, at one time, sponges were thought to be plants — by scientists, no less!  This just goes to show you that scientists are still learning.  As a result, they sometimes change their minds as they get more information.

So what's the purpose of a sponge?  As consumers, what do they consume?  Well, God's great purpose for a sponge is to clean.  They don't just clean our bodies and our kitchen counters when we use them. Sponges actually clean up the ocean! They are some of the most important filter feeders that God made. Do you remember what a filter feeder is? It is an animal that filters its food out of the water. Well, most sponges filter bacteria, algae, filth, and debris out of the water and release clean water back into the ocean. This is a valuable service, because those things cloud the water. In other words, most sponges are living filters that keep the waters clean. They are such dedicated cleaners that a full-grown sponge can filter as much as a bathtub full of water every hour!

Sponges are sessile when they are adults, gluing themselves to a hard surface and staying put. However, when they are larvae, they *do* move around. A young larval sponge can swim, but not strongly enough to overcome the currents. Thus, it is a part of the plankton, floating around until it settles onto a hard surface to which it can glue itself. Sometimes, the tiny larval sponge might settle on the hull of a ship, the back of a crab, or the shell of some other animal. If this happens, the larva will glue itself to the surface and grow into an adult sponge. In this situation, the sponge might end up moving around as an adult, but the sponge really doesn't move around. It is just carried here and there by the ship or animal. So you now know that a sponge isn't always sessile. It is mobile, but only in its larval form.

Sponges grow in both the saltwater of the ocean and in the freshwater of lakes and slowly running rivers. They can be found in both warm waters and in freezing polar waters. They can be deep, deep in the ocean or right near the shore. Wherever they are found, they play a very important role in keeping the waters clean.

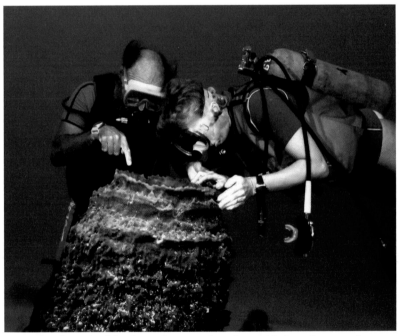

The divers in this photo show you how large this barrel sponge is.

A full-grown sponge may be 2 inches long or 10 feet long. Look at these divers next to a large barrel sponge. Either one could stick his head right inside if he wished to do so. Would you like to do that? It might seem like it would be fun, but you really wouldn't want to do that. You'll find out why in a moment.

# Sponge Anatomy

You have an outer layer to your body; it is called skin. Well, a sponge has an outer layer as well, and it is called the **pinacoderm** (pin ak' uh derm). This outer layer is riddled with holes, or pores.  In fact, sponges are put in phylum **Porifera** (poor if' uh ruh), which means "pore-bearer."  To be scientifically correct, the holes are actually called **ostia** (ahs' tee uh).  What do the ostia do?  Take a guess.  The ostia work like hundreds of tiny doors that open to let water into the sponge.  You see, cells called **collar cells** have little tails that flap back and forth, creating a current that pulls water through the ostia, into the sponge, and out a hole in the top of the sponge.  The water that the ostia let in has oxygen and food particles in it.

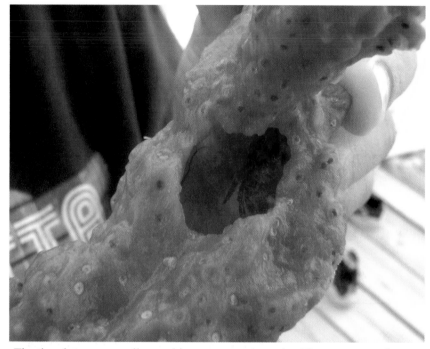

The cells of the sponge absorb the oxygen they need from the water, and the collar cells have tiny collars that trap the food particles carried in by the water.  The food is absorbed by the collar cells and transported to the entire sponge.  Once the food and oxygen are consumed, the now clean water is pushed upward and out the top, where it exits through a large hole called the **osculum** (ahs' kyoo luhm).

The tiny dots you see all over this sponge are the ostia.  They are currently closed, which stops the flow of water through the sponge.  If you look into the osculum closely, you will see a crab inside the sponge.

When small sea creatures find the osculum of a sponge, they'll sometimes swim inside and make the sponge their home.  This works out well for these creatures, because the sponge allows them to hide from predators.  There they live safely protected, eating other creatures that swim inside the osculum.  There are times when a small animal swims inside a small sponge osculum and grows so large while living there that it can never exit again.  It's trapped inside the sponge!  Most sponges have many different kinds of animals and organisms making the sponge their home.  One large sponge was found to have more than *a thousand* organisms living inside it.

# Defenses

Although a sponge can't move once it has attached itself to a surface, there are parts of the sponge that do move.  The cells surrounding the ostia can force the ostia to close, which prevents water from entering.  This keeps bad things, such as poisons or anything else that might harm the sponge, from entering the sponge.  This, then, is one way that the sponge protects itself.

This dead sponge was found washed up on the beach.  Note the large osculum.

Some sponges also have the defense of an offensive odor and disgusting taste.  The odor lets a potential predator know about the taste.  Most animals won't eat these stinky sponges.  But not all sea creatures dislike the taste of a stinky sponge; sea stars, snails, and some fish enjoy a spongy meal now and again.

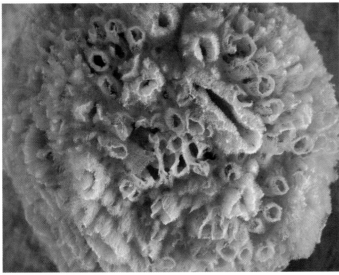
This is the skeleton of a sponge without hard spicules.  It is used for cleaning and artwork.

Another defense that most sponges have is composed of many tiny, sharp spikes that are situated throughout the body.  These spikes, called **spicules** (spik' yoolz), are actually the sponge's skeleton.  They give its body shape and firmness, but also help to discourage other animals from eating the sponge.  Aren't you glad you don't have tiny spikes in your skin?  Predators learn to stay away after getting a mouth full of spicules.  These thorny spikes are why you wouldn't want to climb inside a sponge, even if you could fit!  Not all sponges have hard, sharp spicules, however.  Some sponges are soft, and if you use natural sponges for cleaning or for artwork, those are the ones you are using.  Actually, when you use a natural sponge, you are not using a *living* sponge.  You are using the skeleton that a dead sponge left behind.

*Can you explain what you have learned about sponges in your own words?*

# Sponge Assortment

The variety of sponges in the ocean is truly amazing.  They come in many different shape, size, and color combinations.  They can be red, orange, yellow, green, blue, purple, or a mix of colors.  They might be shaped like long fingers, round barrels, tubes, potatoes, bowls, cups, rocks, or vases.  Sponges growing on coral reefs are so colorful that people often mistake them for coral until they swim up and examine them a little closer.

This student is holding a finger sponge, which looks a lot like a coral, or even a tree branch.

Sponges come in many textures. Many are soft but prickly, while some are even hard and stony. These stony sponges look and feel like small, stony cups. The sponges with which you are most familiar are the ones that look a lot like the sponges we use for cleaning. They are soft sponges with no spicules, and they tend to live in warm waters. Because of their soft, porous body, they soak up water extremely well and have been used for bathing since the time of the Greeks. Roman soldiers also used sponges to pad their heavy helmets and armor. At one time, sponges were taken from the ocean in such huge numbers that some species were on their way to becoming extinct. However, scientists came to the rescue and developed a type of synthetic material that did a great job of soaking up water. This synthetic sponge is the kind you have in your kitchen. Natural sponges are not used very much anymore, but they can still be found in bath shops and in hardware stores for use in creating painting effects. You can also find them in art supply stores.

Have you ever walked along the seashore and found shells with small indentations in them? Those indentations might have been made by **yellow boring sponges**. These sponges grow on clams and other seashells by using acid to bore holes into the shells to which they attach.

Another kind of sponge, the **hermit crab sponge**, attaches to gastropod shells. If a hermit crab takes up residence in the gastropod shell once the gastropod dies, the hermit crab will move the sponge right along with the shell. As the sponge continues to grow, the shell becomes buried in the sponge. This is no problem for the hermit crab, however, because it just moves out of the shell and lives in a chamber in the sponge.

# Making New Sponges

This basket sponge is releasing its sperm into the water to reproduce.

Most sponges are hermaphroditic, having the ability to act as both males and females. When acting as a male, the sponge releases a substance called **sperm** into the water. When a sponge of the same species catches the sperm, it will use the sperm to fertilize its eggs. This will form larvae, which will develop in the sponge for a while. When the larvae are ready, the sponge will release them into the water.

The larval sponge floats in the currents until it settles to the seafloor. If it finds a hard surface upon which to attach itself, it begins transforming into an adult sponge. Depending on the species, the adult sponge will live for one or more years. There are some sponges that are thought to have an unlimited lifespan. They usually die due to being eaten or harmed in some other way.

Another way that sponges can create new sponges is by budding. Do you remember what budding is? It is when a bump appears on the animal and grows. Eventually, the bump separates from the animal and grows into an exact copy of the original animal. In addition, a sponge can grow a whole new sponge from a small piece that breaks off. Once again, this sponge will be an exact copy of the original one. If you have a saltwater aquarium, you can help your sponges become many sponges by breaking off little bits and allowing them to regenerate into new sponges.

*Can you tell someone everything you now know about sponges?*

# Sea Squirts

We're nearing the end of this zoology course, but no book about sea life would be complete without mentioning **sea squirts**. If you've ever seen a rubbery-looking cup with two openings wash up on the beach in a bed of kelp, you've probably seen a sea squirt. These animals can be colorful, and most are usually at least partially transparent, which means you can see through them.

The sea squirt has two openings in its little body. One opening sucks water into the animal, and it's called the **oral siphon**.

This sea squirt is transparent enough for you to see the basket-like sieve that filters food from the water.

The other opening squirts water out of the animal, and it's called the **atrial** (aye' tree uhl) **siphon**. Inside is a little basket-like sieve that traps food. So, you may have guessed that these little rubbery balls are filter feeders. Sometimes you can't even tell that there is an opening in the sea squirt. That's because a sea squirt can draw up the holes in its siphons, sort of like a drawstring closes the opening in a bag. If you picked up a sea squirt and squeezed it, water would squirt out. That's why it's called a sea squirt.

In its larval stage, the sea squirt resembles a tadpole and is therefore sometimes called a **tadpole larva.** Like many sea creatures, then, a sea squirt larva looks nothing like an adult. The larva swims around for a brief period and then attaches itself to something on the seafloor (like a rock). At that point, it transforms into its adult form and usually remains in one place the rest of its life, which generally lasts a few years. Sea squirts are also known as a **tunicates** (too' nih kitz), because the adult form is covered by a leathery structure that can be thought of as a tunic. This tunic supports the animal so that it can stay upright in the water.

# Creation Confirmation

A sea squirt has no brain, no eyes, no arms, nothing that resembles an animal. In fact, like the sponge, the sea squirt used to be considered a plant. Well, you may have a hard time believing this, but there are certain structures in this animal's larval form that make some scientists think that we are actually *related* to sea squirts. Yes, they think that perhaps we were all once sea squirts that evolved into people over hundreds of millions of years. Strange, I know, but here's why they think this. Sea squirts, although they are invertebrates, have something in their bodies similar to the spinal cord that we have inside our backbones. Now, sea squirts don't have backbones, because they are invertebrates. But as I said, a sea squirt larva does have something similar to a spinal cord, so some believe this is an "early form" of a spinal cord. Because of this and a few other incidental things, some scientists mistakenly believe that sea squirts are related to people. "After all," they say, "the reason you look like your parents is because you are related to them. In the same way, sea squirt larvae have some structures that look like human organs because people are related to sea squirts."

The upper car is a Corvette from 1976. The lower one is a Corvette from 2001. They look similar because they come from similar designs, not because they have the same ancestor.

This, of course, is nonsense. Do you know what a Corvette is? It is a kind of car. If you look at all of the Corvettes that have been made over the years, they have a *lot* of similarities. In fact, many of the parts of one model Corvette are nearly identical to the parts in an earlier model. Does this mean that the earlier Corvettes evolved into the later Corvettes and therefore all Corvettes are related? Of course not. Corvettes all look similar because they are all designed and built by the same company. This company uses similar parts and similar designs in all its Corvettes.

In the same way, when God created the animals, He used similar methods for making them, and He used similar organs and body parts that would enable each basic kind of animal to be all that He wanted it to be. Scientists who believe in God can see that the similarities among animals don't mean they share the same ancestor. Instead, some animals look similar to other animals because they share the same Designer.

## Try This!

To get a real feel for how important filter feeders such as sponges and sea squirts are, try this experiment. Collect some pond water in a jar. Get a clipping from a vine or rose and place it in the water. Observe what happens to the water over the next few days and weeks. What do you think would have changed if a filter feeder had been in the water?  Check the "Answers to the Narrative Questions" at the back of the book to find out if you're right.

# Water Worms

Do you like worms?  If you said, "yes," then you aren't like most people.  Most people can't stand the sight or even the thought of worms.  Yet, worms are everywhere.  They are on land, in the sea, and in all freshwater lakes, rivers, and streams.  Let's take a look at a few of the different kinds of worms in creation.  I will concentrate on worms that come from two different phyla: phylum **Annelida** (an' uh lee' duh) and phylum **Platyhelminthes** (plat ee hel min' theez).

# Phylum Annelida

The world is teaming with members of this phylum, which are also called **annelids**.  Not only can you find them in the ground, but you also find them under the waters of ponds, marshes, streams, and the oceans.   From earthworms to leeches, there are annelids everywhere.  We rarely see them because they're usually hidden in dirt, sand, or mud.

An annelid is also called a **segmented worm** because it's divided into many ring-like sections called segments.  If you look closely, you can usually see these little rings around their bodies.  Also, most annelids need to be wet to breathe.  If they can't stay wet, they dry up and die.

# Leeches

Equipped with suckers on both ends, the leech is a fearsome, bloodsucking parasite.  It generally uses its front sucker for sucking blood and its rear sucker for grabbing onto surfaces to help it move.  Leeches haunt freshwater and saltwater habitats.

A hungry leech catches the scent of animals or people in the water.  It swims over and attaches itself by rubbing the sharp teeth in its front sucker against the skin, sort of like the lampreys we studied earlier.  After breaking the skin, it latches on with a powerful suction, feasting on the blood and fluids of the creature.  It'll stay there until it's removed, or until it has had its fill, which won't be for a while, for it can drink up to five times it own weight in blood and fluids.  Once full, it releases itself.  Thankfully, it will be a while before the leech is hungry again.

If you look closely, you can see the rings that separate this leech into different segments. That tells you it's an annelid.

# Bristle Worms

These bristle worms are commonly called "fire worms," because if you get stuck by their bristles, it feels like your skin is on fire.

Bristle worms have bundles of tiny hairs on each segment of the body. They actually look like a cross between a caterpillar and a centipede. The hair bundles, which often look like legs, can be used to help the worms swim through the water, as shovels to dig under the sand or mud, or as hooks to hold on to objects on the seafloor. Many bristle worms have poison bristles. The bristles will break off in the skin of a predator that tries to pick up the animal, stinging terribly.

A bristle worm's head has tentacles, which might look like thin fingers or a crown of feathers. Most have eyes that can tell dark from light, but they cannot actually see things. It is believed that a few can make out vague shapes. To enhance their senses, some bristle worms have a pair of feelers at the ends of their bodies.

These worms can be teeny-tiny or several feet long. Most are scavengers, but some are accomplished predators, eating fish and coral. Others munch on algae. They are typically found hiding in reefs and rocky shores. They prowl along the seafloor or the bottom of a tide pool, seeking something to devour.

On the seashore, you may find two kinds of bristle worms: **ragworms** and **lugworms**. Ragworms live in burrows they make in the sand or mud. If enough sand or mud has collected in the bottom of a tide pool, you can usually find ragworms there. When the tide is out, the ragworm stays hidden under the sand. When the tide comes in, the ragworm comes out from its hole to feed on seaweed and dead animals. It has strong teeth, which can bite into your skin, though it's not a dangerous animal. It wiggles through the wet mud or sand like a snake or swims through the water by paddling its bristles.

Have you ever noticed little mounds of sand or mud on the shore? These are made by the lugworm, which burrows into the sand. The lugworm hides in a U-shaped burrow with two holes. It wiggles its body to create a current that brings water into its burrow. Why do you think it might need water? It takes the oxygen out of the water and uses it to breathe. It swallows sand or mud, pulling out anything edible and passing the remainder out as waste. This waste sand or mud ends up in a pile at the top of the hole where its tail end sticks out. Next time you are at the beach, search for these little piles! Inside is a lugworm eating sand.

Other types of bristle worms build tubes around themselves. Some use sand particles and bits of shells glued together with mucus to make their tubes. Others use chemicals they collect from the ocean to produce limestone from which they make their tubes. Not surprisingly, these worms are often called **tubeworms**. They have fan-like tentacles that are coated with mucus. When they are feeding, they extend their tentacles in the water so that food particles floating in the water can be caught in the mucus. Sometimes tubeworms live in groups, and when all their feathery tentacles are out, they look like a bouquet of flowers.

This tubeworm is called a Christmas tree worm, because its tentacles look like two little Christmas trees.

If fish or other predators swim by, a tubeworm immediately pulls its tentacles and the rest of its body into the safety of its tube. This makes them difficult to find. If you were to swim past in your scuba gear, for example, any bristle worms in the area would most likely pull into their tubes so quickly that you might think the seabed was dead and lifeless.

These giant tubeworms at the bottom of the ocean work together with bacteria so that they both can survive in the hot, chemical-filled water.

In the freezing, dark night of the deep ocean, where scientists at one time thought no life could survive, some of the most amazing creatures can be found. Giant tubeworms, 10 feet long and as thick as your arm, grow in great clusters near deep-sea vents. These vents at the bottom of the ocean spew blisteringly hot water and chemicals that are toxic to most animals. Most animals would not be able to survive at such temperatures and near these chemicals, but these huge worms inside hard tubes can.

How do the tubeworms do it? They have a symbiotic relationship with bacteria. Do you remember what symbiosis is? It is when two or more types of creatures work together so that both can live. In this symbiotic relationship, the tubeworm collects chemicals that are spewing out of the deep-sea vent and gives them to the bacteria. In return, the bacteria use those chemicals to make energy for themselves and for the tubeworm.

*All of the worms you have learned about so far are annelids.*
*Tell someone everything you remember about them.*

# Flatworms

Flatworms are…you guessed it…flat…very, very flat. They belong in phylum Platyhelminthes. There are many different kinds of flatworms, and some swim around in freshwater while others are found in saltwater. They don't have ring-like sections on their bodies like annelids have. Some flatworms live inside the bodies of other animals as parasites, robbing those animals of nutrients and body tissue. Others, which don't

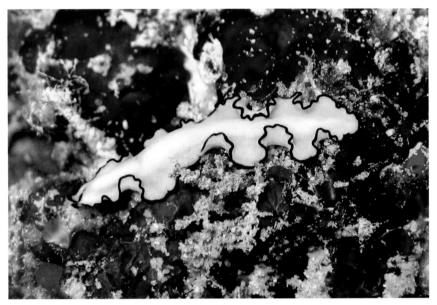
This is a free-living, marine flatworm.

live inside the bodies of other animals, are called free-living flatworms. Many free-living flatworms live in the ocean, usually near the shore or in tide pools or reefs. Some live in freshwater, and others live in damp dirt.

The free-living flatworm has simple eyes on the front of its head, which is usually wider than its back end. Some also have feelers on their heads. On the underside of a free-living flatworm's body, there are tiny hairs, called cilia. The cilia wave back and forth to help the worm move along the seafloor. Most free-living flatworms are hunters, eating animals that are smaller than themselves. Like the cnidarians you learned about earlier, there is only one opening in a flatworm. As a result, they both take in food and eliminate waste through the mouth.

Another thing flatworms do is regenerate. We've learned a lot about regeneration, haven't we? Well, a flatworm sometimes intentionally splits itself in half to form another flatworm. Of course, when this happens, the new flatworm that forms is an exact copy, a clone, of the original.

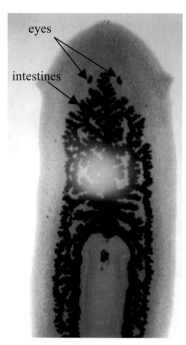

eyes

intestines

This is a magnified image of the front end of a planarian.

As I told you before, some flatworms live in freshwater and others in saltwater. Many freshwater flatworms are called **planarians** (pluh nar' ee uhnz). They live all over the world, hiding in rocks and weeds. They are usually not noticed because they are about the size of the tip of your finger, or smaller, and they usually emerge at night.

Like many worms, they have eyes, but really cannot see things. Instead, their eyes allow planarians to see the difference between light and dark. Their intestines, which digest their food, cover the entire length of the body.

If you feel really adventurous, you can actually buy live planarians so that you can study them yourself! You will need a magnifying glass to study them well, however, because they are very small. If you are interested in doing this, the course website that is listed in the introduction to this book has links to places from which you can purchase live planarians.

Marine flatworms are much larger than those that live in freshwater, and they often have ruffled or curly edges around their bodies. Because they are so brightly colored, marine flatworms can be easily mistaken for the beautiful nudibranchs you studied previously. Why do you think God made these flatworms so that they look a lot like nudibranchs?

Well, if you remember, a nudibranch's bright colors warn predators that it is not very tasty and perhaps even toxic. Once a predator has taken a small bite out of a nudibranch, it remembers never to touch that colored creature again. So any creatures that have had the distasteful experience of trying to eat a nudibranch will also stay away from the marine flatworm of the same color. This kind of defense is called **mimicry**, because a marine flatworm mimics a nudibranch so as to fool predators into thinking it is not a tasty meal.

The course website I told you about in the introduction to this book has a very interesting link that takes you to a website with pictures of nudibranchs and the marine flatworms that mimic them. It is very interesting to see how good these marine flatworms are at mimicry.

By quickly beating the cilia on the undersides, of their bellies, marine flatworms are able to move along the seafloor much more quickly than nudibranchs. They are also much thinner than nudibranchs. In fact, if you picked up a marine flatworm, it might just break apart in your hands. Of course, if it did break apart, the pieces could grow back into new flatworms!

*Tell someone what you have learned about flatworms.*

# Tiny Tales

We are coming to the end of this course, but I must discuss one last group of animals with you. They are often neglected in the search for water animals, because they are so small that a microscope is needed in order to see them.

If you were to take a sample of water from a pond or the sea on a warm summer evening and study the contents under a microscope, you would find an astounding number of living things. Many, many different kinds of life would be swimming, paddling, and lurching about in your tiny sample of water. Many of these microscopic creatures would not be animals at all. Remember, to be an animal, a creature has to be multicellular. Many of the microscopic creatures in creation are unicellular. Thus, they are not

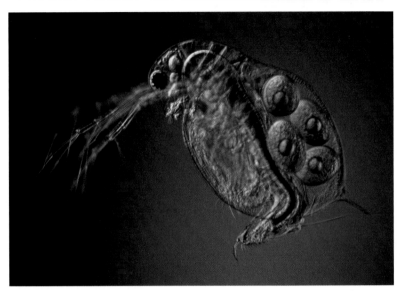

This is a microscopic freshwater animal from genus *Daphnia*. It is actually a tiny crustacean.

animals. Nevertheless, many microscopic creatures are multicellular. They are the **microscopic animals**.

So what kinds of microscopic animals might you find in a sample of pond water? Well, you might find rotifers, water bears, microscopic worms, and microscopic crustaceans, such as the one pictured above. In a sample from the sea, you might also find the microscopic larvae of many of the animals you have studied in this course. In one pail of water, you could easily find thousands of these little creatures, but you would need a microscope to actually see that they were there. Little did you know that when you swam in a pond or the ocean you were swimming in the midst of so many tiny animals! It sort of makes you want to take a warm, soapy bath after swimming, doesn't it?

Let's take a quick peek at two of the many different kinds of microscopic animals that you might find in a pond or lake. If you ever get a microscope, you might want to take some samples of pond or lake water to see if you can find any of these interesting creatures.

# Rotifers

Way back in the 1600s, when microscopes were newly invented, a tiny creature was discovered. Around what looked like a mouth were tiny little hairs (cilia) that flapped back and forth in such a way as to resemble a wheel turning.  Because of this, these animals were named "wheel animals" or **rotifers** (roh' tuh furz).  The cilia that give them their name help rotifers to move, but they are mostly used to eat.  When food (basically anything smaller than the mouth) nears the animal, the cilia create a current that causes the food to be drawn into the creature's mouth.  The food is digested inside the animal and waste is moved out through an exit hole near the bottom of its little body.

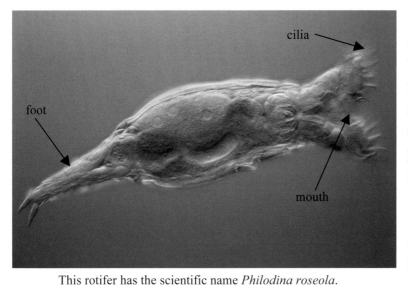
This rotifer has the scientific name *Philodina roseola.*

Most rotifers also have a tail-like projection at the end of the body.  This is actually the rotifer's foot.  At the tip of the foot are claws that are able to grip onto surfaces.  Some rotifers can also attach themselves to plants and objects by using a sticky substance that comes out of the foot.

Most rotifers are less than 2 mm (about one-twelfth of an inch) in size, and they are typically box-shaped or worm-shaped.  The rotifer in the picture is one of the worm-shaped rotifers.  They all have at least one eye that can tell light from dark.

Have you ever noticed dust particles flying about in the air?  Well, don't be surprised if there are some rotifer eggs in that dust.  If eggs are washed ashore, they dry out and get caught in the wind.  Once they land in water, they can hatch.  Because the eggs travel so much, even small puddles of water in your backyard may have rotifers; they can be found anywhere that water is standing!  Unbelievably, if the water dries up, the rotifer can still survive!  It does this by going into a state called a **tun**.  In this state, the rotifer dries up into a little case and stays there until it's wet enough to return to rotifer life.

# Tardigrades

Do you realize that you may have bears living in your backyard?  It's true.  You probably have hundreds or thousands of bears ambling around in the puddles of water in your yard.  Actually, these little "water bears" only look like bears under a microscope, and the biggest they will grow is as big as the head of a pin.  Even so, the water bear is a cute little critter, with a short, stubby body that ambles slowly along on its eight bear-like legs, which have bear-like claws on the end of each foot.

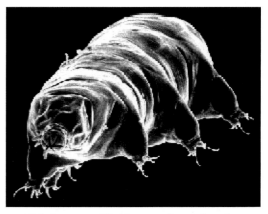
This is a magnified image of a tardigrade.

These water bears walk so slowly that scientists call them "tardy," **tardigrades** (tar' duh graydz), that is. We generally use the word "tardy" to mean someone is late. Well, in Latin, "*tardus*" means "slow" and "*gradus*" means "walker." Water bears, then, are slow walkers.

Even though a tardigrade looks like a bear, it is definitely not a bear. For one thing, a bear doesn't have tiny needles on its mouth! Yikes. For what are the needles used? They are used for eating, of course. A tardigrade uses these needles to pierce into creatures and suck the juices out of them.

Like the eggs of rotifers, tardigrade eggs are light enough to be blown around on the wind. Thus, like rotifers, they can be found in most bodies of water. In fact, because a tardigrade can also form a tun, just like a rotifer, it can survive in bodies of water that sometimes dry up. When the water dries out, a tardigrade just becomes a tun and waits for the rain to come. It can stay like this for months or even years! When it rains again, the bear comes out of "hibernation," ambling about your lawn like it's the king of the forest.

*Put in your own words all that you have learned about microscopic animals.*

# Summing It Up

Well, you have come to the end of your studies. You have learned a lot about aquatic life, but there is a lot more to learn! If you found this course interesting, you might want to spend even more time learning about the wonders of God's aquatic creatures. One day, perhaps you will become a scientist who will help us know even more about the mysterious creatures of the sea. In whatever you do, remember that God is the Creator of all that is around you. Look up and glorify Him when you study what He has made.

# What Do You Remember?

How do scientists decide if a strange-looking creature is an animal? What do sponges do for the water environment around them? What are the ostia in a sponge? What is the osculum of a sponge? Why do some animals have features that are similar to those of other animals? For what are the bristle worm's bristles used? How do leeches eat? Why do some marine flatworms look like nudibranchs? How do rotifers eat? Why are they called rotifers? What are water bears called? Why can water bears and rotifers be found in so many bodies of water?

# Notebook Activities

After writing down the fun facts you learned in this lesson, make at least three pages for your notebook. One should be about sponges, one should be about water worms, and one should be about microscopic animals. The *Zoology 2 Notebooking Journal* has pages for these activities.

# Ocean Box

You will want to make sponges and worms for your ocean box. Colorful clay will be perfect for sponges and flatworms. To make marine tubeworms, you can use white straws. Cut them in half or quarters. Stick a piece of clay into the tip of each straw, and then arrange feathers around the tip to represent the tentacles of the tubeworms. Insert them into a chunk of clay so that they can stand upright.

To form sponges, roll a piece of clay into a thin cylinder. Then, insert a pencil or some other thin object into the center to form a hole. Remove the pencil and use its tip to make ostia all over the outside of the sponge.

# Experiment

We have discussed that the world is mostly water, but most of that water is salty. Saltwater, though great for many animals, is not potable (poh' tuh bul — drinkable) for humans, land animals, freshwater animals or even the crops we grow for food. So despite the fact that most water on earth is saltwater, everyone must have freshwater to live.

Can you think of any places on earth that might have a more difficult time getting freshwater than others? The desert is one such place. In the country of Saudi Arabia, water costs more than gasoline! That's because they get a lot of their water from the sea, and sea water must have the salt removed before it is drinkable. This process is called **desalination** (dee sal' uh nay' shun). There are different ways to desalinate saltwater. You will explore one of them in this experiment.

**You will need:**
- A "Scientific Speculation Sheet"
- Two water bottles
- Black paint
- One cup of water mixed with one tablespoon of salt
- Plastic tubing
- Clay
- A nice sunny window
- A stack of books

1. Paint one water bottle black.

2. Taste your saltwater mixture.  **NOTE:** You should ***never*** taste *anything* in an experiment unless someone who knows exactly what the experiment involves tells you to do so.  Some experiments can produce things that are poisonous to drink.  There is nothing poisonous about a tiny taste of saltwater, however, which is why I am telling you to taste the saltwater.

3. When the paint is dry, pour your saltwater mixture into the black bottle.

4. Surround one end of the plastic tube with a thick glob of clay, but don't allow the clay to stop up the tube.  Let the end of the tube stick a little way out of the glob of clay.

5. Stick the clay and the end of the tube in the mouth of the black bottle, making an airtight seal around the top of the bottle and the tube.

6. Connect the other end of the tube to the other empty bottle in exactly the same way.

7. Place the bottles in a sunny window or near a warm lamp, setting the black bottle on blocks or books to elevate it above the clear bottle.

8. What do you predict will happen?  Write your prediction down on a "Scientific Speculation Sheet."

9. Check back the next day to see what happened.  You should see a little bit of water in the clear bottle.

10. Keep checking each day until you have enough water in the clear bottle to taste.  Once you have enough, taste the water.  Does it taste salty?

What happened?  The black paint on the bottle that contained saltwater absorbed a lot of the light from the sun or the lamp.  This caused the saltwater to get hot, and some of it evaporated.  That means it turned into water vapor, which floats in the air.  When the water evaporated, however, it could not take the salt with it, so the salt was left behind.  Some of the water vapor made its way into the tube, and once it cooled, it became liquid again, dripping down into the clear bottle.  When you tasted the water in the clear bottle, you tasted no salt, because there was none there.

This is one way freshwater is made from saltwater.  The process is called **distillation**, and it involves evaporating water from saltwater and then collecting the water vapor and cooling it down so that it becomes liquid again.

# Answers to the Narrative Questions

Your child should not be expected to know the answer to every question. These questions are designed to jog the child's memory and help him put the concepts into his own words. *The questions are highlighted in bold and italic type*. The answers are in plain type.

## Lesson 1

***What are nektonic creatures?*** They are animals that can swim. ***What are benthic creatures?*** They are animals that do not swim but move across the bottom of a body of water. ***What are plankton?*** They are animals that drift with the currents. ***Where can zooplankton be found at night?*** They can be found near the surface of the water. ***Why are plankton important to all sea life?*** They are food for many sea creatures. ***What are filter feeders?*** They are animals that filter their food out of the water. ***Can you name the five oceans in the world?*** The five oceans are the Pacific Ocean, the Atlantic Ocean, the Indian Ocean, the Southern Ocean and the Arctic Ocean. ***What are seas?*** Seas are smaller than oceans, but they are made up of saltwater because they are connected to oceans. ***What are estuaries?*** They are the places where a river meets an ocean or sea. ***Beginning from the shore out to the deep, what are the four zones of the ocean floor?*** In order, they are the continental shelf, the continental slope, the continental rise, and the abyssal plain. ***From the surface of the ocean to the deep, what are the three zones in which aquatic creatures live?*** They are the sunlit zone, the twilight zone, and the midnight zone. ***What are the circular currents called?*** They are called gyres. ***What are the currents caused by temperature and salt levels called?*** They are called thermohaline currents. ***What causes the tides?*** The moon causes the tides.

## Lesson 2

***Answer to the "Try This" activity on page 32***: The saltwater took longer to freeze or did not freeze at all. This is because salt lowers the freezing temperature of water. ***How does a cetacean move its tail to propel itself through the water?*** It moves its tail up and down. ***Which is the most important sense for a whale: smelling, hearing, or seeing?*** Hearing is the most important sense for a whale. ***What must a calf do as soon as it is born?*** It must get to the surface to breathe. ***How does the mother help it do this?*** The mother guides the calf to the surface. ***Why must a whale have a blowhole?*** The blowhole allows it to breathe when it reaches the surface of the water. ***Where do most whales spend the summer and winter? Why?*** They mostly spend the summer in the polar regions, because that's where there is a good food supply. They spend the winter in the more tropical regions in order to have their young. ***What is breaching?*** It is when a whale leaps into the air and then purposefully flops down on the water with a splash. ***What is lobtailing?*** It is when a whale faces downward in the water with only its fluke sticking out and slaps the water with its fluke. ***What is spyhopping?*** It is when a whale is in an upright position with its head straight up and out of the water. ***What is logging?*** It is when a whale swims slowly on the surface of the ocean with very little movement. ***Why did whalers want to kill whales?*** They wanted the whales' blubber, which could be used to make many useful things. ***Which two kinds of whales did whalers really like?*** They liked right whales and sperm whales. ***How are toothed whales different from baleen whales?*** Toothed whales have teeth in their mouths. ***What kind of whale has a "horn" like a unicorn?*** The narwhal has a horn like a unicorn. ***Name a difference between dolphins and porpoises.*** Porpoises are smaller than dolphins, dolphins have beaks, while porpoises do not, and a dolphin's dorsal fin curves, while a porpoise's dorsal fin does not. ***What is the largest animal on earth?*** The blue whale is the largest animal on earth.

## Lesson 3

***Answers to the identification activity on page 47:*** A. seal  B. fur seal  C. sea lion  D. sea lion  E. fur seal  F. seal
***What is the main difference between a true seal and a sea lion?*** A true seal has no external ear flaps, while a sea lion does. ***What are the differences between a fur seal and a sea lion?*** Sea lions are usually larger than fur seals, have more

rounded snouts than fur seals, have shorter flippers (compared to the body) than fur seals, and have rougher and shorter fur than fur seals. ***What is a haul out?*** It is when a pinniped pulls its body out of the water and onto land. ***What is a rookery?*** It is a pinniped's breeding ground. ***What are some dangers to pinnipeds?*** Polar bears, killers whales, and sharks love to eat pinnipeds. A pinniped pup can be caught in a bad storm and washed away from its rookery. A pinniped can get tangled in fishing line or hooks can get caught in their skin. During times when food is not in good supply, pinniped pups don't get enough to eat when they nurse from their undernourished mothers. ***How does a walrus differ from other pinnipeds?*** It has tusks. ***What does the walrus family name, Odobenidae, mean?*** It means "one that walks with teeth." ***What is the main difference between a manatee and dugong?*** A manatee's tail is shaped like a beaver's tail (rounded), while a dugong's tail is shaped like a whale's tail (more triangular). ***What temperature of water do manatees like?*** They like warm water. ***What do manatees do when they meet one another?*** They "kiss" each other. ***Why do manatees need to stay in shallow water?*** Their food supply is in shallow water, and they sleep at the bottom of a body of water, but every few minutes, they must rise to the surface for a breath and then drop back down to the bottom. This is easiest to do in shallow water. ***Why is this dangerous for them?*** This makes them a target for boats to hit.

## Lesson 4

***What is the name we use to mean both amphibians and reptiles?*** We call them "herps." ***Name some of the differences between mammals and reptiles.*** Reptiles are oviparous and ectothermic; mammals are not. ***How are sea turtles different from land turtles?*** A sea turtle has flippers instead of claws, it cannot pull itself completely into its shell, and its shell is flatter than that of a land turtle. ***What is the top part of a turtle's shell called?*** It is called the carapace. ***What term do we use to describe hibernation in herps?*** We call it brumation. ***What are some of the dangers sea turtles face?*** Sea turtle hatchlings can head in the wrong direction and never make it to sea, they can be eaten on their way out to sea, and jellyfish-eating turtles can eat plastic bags and choke to death. ***What makes a sea snake different from other snakes?*** They cannot survive on land, they don't have scutes, they have a paddle-shaped tail, and they have valved nostrils. ***What are the two kinds of snake venom?*** Snake venom can be made of hemotoxin or neurotoxin. ***How are reptiles different from amphibians?*** Amphibian eggs need to stay in water, while reptile eggs do not. Amphibian eggs are soft, while reptile eggs are harder. Amphibians look nothing like their parents when they hatch, while reptiles look like miniature versions of their parents when they hatch. Reptiles have dry scales made of keratin, while amphibian skin has no scales and must stay moist. ***What is the difference between most amphibians and aquatic amphibians?*** Most amphibians leave the water for long periods of time when they are adults; aquatic amphibians do not.

## Lesson 5

***Who found the first fossil of a giant "sea monster?"*** Mary and Joseph Anning found the first sea monster. Mary is the one who is really remembered for it. ***What four kinds of large sea reptiles did we discuss?*** We discussed nothosaurs, mosasaurs, ichthyosaurs, and plesiosaurs. ***Which of those four might not have spent all of its time in the sea?*** The nothosaur might have spent part of its life on land. ***Which two animals is a plesiosaur like, and how is it like them?*** It is like a turtle because it has four flippers it uses to swim and gastralia. It is like a snake because of its long neck. ***What did the plesiosaur eat to aid in chewing its food?*** It ate large stones that we call gastroliths. ***How do we know what an ichthyosaur looks like?*** A fossil was found that has the outline of the body. ***How is an ichthyosaur different from a fish?*** Its flippers enclose arm, hand, and finger bones. A fish does not have such bones. In the same way, the backbone is not like a fish's backbone; it is like a reptile's backbone. ***Which of the four types of giant marine reptiles was shaped a lot like a snake and had tiny flippers?*** The mosasaur was shaped like a snake and had tiny flippers. ***What is one explanation for why we have so many sea creature fossils all over the earth?*** The sea creatures were caught in sediments that churned in the sea as a result of the global Flood.

## Lesson 6

***What makes a fish a fish?*** A fish has fins for swimming and gills for breathing. ***What fish shape is designed for fast swimming?*** The fusiform shape is designed for fast swimming. ***What did God give fish to help them stay buoyant in the***

*water?* He gave them a swim bladder. *How many nostrils does a fish have?* A fish has four nostrils. *Name two defenses that a fish might have.* Some fish swim quickly, some use camouflage, some use advertising, and some swim in schools. *What does "osteichthyes" mean?* It means bony fish. *What does a fish's lateral line do?* It senses tiny vibrations that travel through the water. *What is spawning?* It is when a female fish lays eggs and then a male fish fertilizes them. *Name a fish that makes a long journey in order to reproduce.* A salmon travels long distances from the ocean to the freshwater stream in which it hatched. *What are the five stages of a typical fish's development?* The five stages are: the egg stage, the larval stage, the postlarval stage, the juvenile stage, and the adult stage.

## Lesson 7

*What does "Chondrichthyes" mean?* It means cartilaginous fish. *What are the scales of sharks and rays like?* They are like teeth. *How are the fins of bony fish and sharks different?* Bony fish have thin, almost transparent fins, while sharks have thick and rubbery fins. *Why do sharks and rays sink when they are not swimming?* They do not have swim bladders. *What is the difference between a manta ray and a stingray?* A stingray spends a lot of its time on the ocean floor, while a manta ray spends most of its time swimming. *How can you tell the size of a shark by its teeth?* Every inch of tooth equals 10 feet of shark length. *How do the ampullae of Lorenzini help a shark?* They detect the electrical signals being given off by the animals in the water, which helps it hunt. *If you are swimming in the ocean and see a shark, what should you do?* Stay calm and don't make rapid movements. Swim calmly and in rhythm toward the land or boat. Keep the shark always in sight. If all else fails, fight the shark with your fists and legs, kicking it in the head and nostrils. *What does "Agnatha" mean?* It means without jaw. *What does "anadromous" mean?* It means that an animal is born in freshwater but lives its adult life in saltwater. *What are some differences between the way a hagfish feeds and the way a lamprey feeds?* Lampreys tend to feed on live things, while hagfish tend to feed on dead things. Also, a lamprey sucks fluids out of its prey's body, while a hagfish tears chunks from its prey. *What do almost all fish have that the hagfish doesn't have?* The hagfish has no backbone.

## Lesson 8

*Picture on page 127:* The lobster on the right is female; the one on the left is male. *Picture on page 128:* The crab on the top is female; the one on the bottom is male. *What does the word "arthropod" mean?* It means jointed foot. *What is an exoskeleton?* It is the armor that covers a crustacean. *How does a crustacean molt?* To molt, an arthropod produces chemicals that weaken its exoskeleton. It then takes in water so that it begins to swell. At the same time, it starts growing a new, flexible exoskeleton under the old, weakened one. When the old exoskeleton gets weak enough, the arthropod breaks out of it. Then, the new exoskeleton hardens, and the arthropod gets rid of all the excess water. *How do antennae help crustaceans?* They give the crustacean strong senses of touch and taste. *What are maxillipeds?* They are small mouthparts that a crustacean uses to hold, touch, and taste food. *What are chelipeds?* They are a crustacean's claws. *What are some of the uses of swimmerets?* They act as paddles that help propel a crustacean through the water, some crustaceans use them to push water over the gills they have in their abdomen, and they are also used by the females to hold eggs. *How long can a lobster live?* A lobster can live to be a hundred years old. *How are crabs different from lobsters?* They don't have antennae or antennules, and they walk sideways. *What is the symbiotic relationship between shrimp and fish?* Some shrimp clean the teeth of some fish. *What does it mean to be a keystone species?* If a keystone species were to die, it would cause a lot of animals to die out as well. *Which crustacean is a keystone species?* Krill are a keystone species. *Where do barnacles live?* They live attached to anything hard that spends some time submerged in water. *Where do horseshoe crabs lay their eggs?* They lay their eggs in holes that they dig in the sand on the shore. *What kind of eyes did trilobites have?* They had compound eyes made out of rock crystals.

## Lesson 9

*What is the main difference between bivalves and gastropods?* Bivalves have a shell that has two halves; a gastropod shell is one piece. *How do bivalves filter feed and breathe?* They have siphon tubes. One siphon pulls water over the gills so the bivalve can breathe. Also, food particles that are floating in the water are removed so that the bivalve can eat. The other siphon pushes the water back out of the bivalve. *Where are live clams found on shore?* They are found buried in the sand. *How can you tell the age of a clam?* You can count the rings on its shell. Each ring means a year of growth. *Which bivalves cling to rocks and other surfaces?* Mussels and oysters are clinging bivalves. *How and when do they find food?* They filter food from the water whenever the water covers them. *Where are pearls found and how are they formed?* They are found in oysters, mussels, and clams. They form when a piece of dirt, sand, or something gets in the animal, and the animal forms a shell around it. *How do scallops swim?* A scallop opens its shell and then closes it quickly, which pushes it through the water. *What does the term "gastropod" mean?* It means stomach foot. *What is a radula?* It is a jagged organ with many denticles that a mollusk uses to get food. *What is an operculum?* It is a "door" that can be used to close a gastropod's shell. *What kind of gastropod makes a shell that has a wide, pearly-colored lip that flares outward?* The conch makes such a shell. *What kind of gastropod has a shell with several large holes?* Abalone shells have holes. *What kind of gastropod has no shell?* A nudibranch has no shell. *What do some nudibranchs do with the stingers of the sea anemones that they eat?* They use them to defend themselves.

## Lesson 10

*What does "cephalopod" mean?* It means head foot. *What are the four different kinds of animals in the cephalapod group?* They are the octopus, the squid, the nautilus, and the cuttlefish. *How do cephalopods swim?* They swim by jet propulsion. *What do cephalopods usually eat?* They eat mostly fish and crabs. *What kind of mouth do cephalopods have?* Cephalopods have beaks. *What are some of the defenses that cephalopods have?* They have camouflage, ink, and excellent eyesight. *What is the internal shell of the cuttlefish called?* It is called a cuttlebone. *How many arms do cuttlefish and squids have?* They have ten arms. *What does a squid usually do after it mates or lays eggs?* It usually dies. *How many arms does an octopus have?* It has eight arms. *Why do scientists think octopuses are intelligent?* They have large brains and good problem-solving skills. *What is different about the nautilus compared to other cephalopods?* It has a lot more arms and creates an external shell. *How does the nautilus move up and down in the water?* It sucks in gas through a tube, which makes water enter the chambers, making the nautilus heavier. This makes the nautilus sink. When it forces gas out of the tube, that pushes water out of the shell, making the nautilus lighter. This makes the nautilus rise in the water. *Describe a chiton.* It is a mollusk that forms an oval-shaped shell with overlapping plates. *Which land animal is it like and why?* It is like a pill bug. *How is it like a gastropod?* It has a single shell that suctions to rocks. *Where might you find a chiton during the day?* You can find one hiding under rock ledges during the day.

## Lesson 11

*What is the name of the phylum that includes sea stars, sand dollars, sea urchins, and sea cucumbers?* It is called phylum Echinodermata. *What does that word mean?* It means spiny skin. *What is special about these animals' feet?* They are tubular strands with suction cups on the end. *Explain how sea stars eat.* A sea star's stomach goes out of its mouth, engulfs its prey, and starts digesting it. *What is a sea star's favorite food?* Its favorite food is bivalves. *What did clam fishermen once do to keep sea stars from eating the clams?* They cut them in half and threw them back into the ocean. *Why did this not work?* It didn't work because sea stars regenerate. *Why are brittle stars considered brittle?* They drop their arms to escape predators, which makes them seem brittle. *How do they move across the ocean floor?* They use muscles in their arms to scuttle rapidly about. *How is this different from sea stars?* Most sea stars use their tube feet to move slowly across the ocean floor. *What are sea urchin's teeth called?* They make up a system called Aristotle's lantern. *What animal really likes to eat sea urchins?* Sea otters love sea urchins. *Explain how a sand dollar eats.* Sand dollars filter sand and water, catching plankton and debris on their spines. Then, they use cilia to move the food into a food groove, and then the food travels down the food groove to the mouth. *How does a sea cucumber defend itself?* It tosses out long sticky threads that are actually made up of its own internal organs. *In what two ways do sea cucumbers eat?* Some eat sand and absorb any food in the sand, and others filter food out of the water.

## Lesson 12

*What does "cnidarian" mean?* It means nettle. *Where is the mouth on a cnidarian?* It is in the center of its tentacles. *Explain how nematocysts work.* The poison arrow of a nematocyst is in a closed capsule until it is triggered. Once triggered, the capsule opens, and the arrow pops out and pierces the skin of the prey. The thread then shoots into the body through the arrow and injects poison into the prey. *What is the difference between a polyp and a medusa?* In a medusa, the tentacles dangle down, while in a polyp, they wave above the animal. *Why are jellyfish considered plankton?* They cannot swim strongly enough to overcome the force of the currents. *What is special about the box jelly?* Its poison is strong enough to kill a person. *What is special about the man-o-war?* It is actually a colony made up of many individuals. *How do sea anemones and corals differ from jellyfish?* Sea anemones and corals are polyps that tend to stay in one place, while jellyfish are mobile medusae. *How do corals differ from sea anemones?* Many corals make a stony home in which they live. *Where do corals grow and why?* They grow in warm waters that get sunlight, because their symbiotic zooxanthellae need light, and they need warm water to build their outer skeletons. *Explain how corals are dependent on algae.* Corals get nutrients from the algae, and the algae help them make their outer skeletons faster. *What is the difference between stony coral and soft coral?* Stony corals make a hard, outer skeleton, while soft corals make tiny limestone spines that are embedded in their soft skin.

## Lesson 13

*"Try This" on pages 208-209:* With a filter feeder, the water would have stayed much cleaner, because the filter feeder would have eaten the debris that formed. *How do scientists decide if a strange-looking creature is an animal?* If a creature is multicellular and consumes food, it is an animal. *What do sponges do for the water environment around them?* They clean the water. *What are the ostia in a sponge?* They are the tiny holes that allow water into the sponge. *What is the osculum of a sponge?* It is the big hole in a sponge through which water exits. *Why do some animals have features that are similar to those of other animals?* Some animals resemble others because they were made by the same Designer, God. *For what are the bristle worm's bristles used?* They are used to help the worm swim through the water, as shovels to dig under the sand or mud, or as hooks to hold on to objects on the seafloor. *How do leeches eat?* They suck the blood of a living creature. *Why do some marine flatworms look like nudibranchs?* This mimicry helps them avoid predators, because the predators think they are bad-tasting nudibranchs. *How do rotifers get around?* When food nears the animal, the cilia create a current that causes the food to be drawn into the creature's mouth. *Why are they called rotifers?* Their cilia move in such a way as to resemble a wheel turning. *What are water bears called?* They are called tardigrades. *Why can water bears and rotifers be found in so many bodies of water?* Their eggs are light enough to blow in the wind, so they can land in any body of water and hatch.

# Photograph and Illustration Credits

**Photos and illustrations from www.clipart.com:** 2, 3 (top), 11, 25, 44 (top), 47 (top left, middle right), 73, 87 (top), 88, 89 (2nd, 3rd, and 4th from left on top), 99 (top), 110, 112 (bottom), 136 (bottom), 148 (top), 176, 194

**Illustrations by Megan Whitaker:** 7, 9, 12, 22, 30 (bottom), 35, 74–77, 79, 81, 82 (top), 90, 118, 123, 162 (top), 186

**Photos by Jeannie K. Fulbright:** 16 (bottom), 17, 18, 39, 40, 53, 54, 57, 71, 101, 102, 105, 108 (bottom), 111 (top), 113, 120, 124 (top), 130, 131, 135, 139, 140 (top two), 141 (both middle), 143, 146 (bottom), 150 (bottom), 153 (both bottom), 154 (all), 155 (top), 157, 158, 169-172 (all), 177 (top), 179 (both), 184. 200, 204, 217, 218

**Photos and illustrations by Dr. Jay L. Wile:** 28 (bottom), 80, 92 (bottom), 94 (both), 128

**Photos by Kathleen J. Wile:** 21, 87 (bottom), 89 (top right), 93 (bottom), 213

**Photos by John Skipper:** 89 (bottom), 124 (bottom), 127

**Photo copyright © photographer/oceanwideimages.com (photographer in parentheses:** 4 (Gary Bell), 14 (Rudie Kuiter), 23 (Gary Bell), 26 (Lin Sutherland), 29 (Bill Boyce), 34 (Lin Sutherland), 38 (Lin Sutherland), 43 (Chris & Monique Fallows), 51 (Bob Halstead), 55 (Gary Bell), 59 (Gary Bell), 61 (Gary Bell – top), 64 (Gary Bell), 65 (Gary Bell), 98 (Rudie Kuiter), 99 (Rudie Kuiter – middle and bottom), 107 (Rudie Kuiter), 108 (Rudie Kuiter – middle), 109 (Rudie Kuiter – top, Gary Bell – bottom), 112 (Chris & Monique Fallows – top), 114 (Rudie Kuiter – middle and bottom), 115 (Rudie Kuiter – top, Gary Bell – bottom), 174 (Gary Bell – bottom), 190 (Gary Bell), 191 (Gary Bell)

**Photos from www.dreamstime.com: (copyright © holder in parentheses):** 46 (David Crehner – bottom), 63 (Daniel Gustavsson – top), 67 (Thomas Mounsey – bottom), 68 (Radu Razvan), 108 (Bartlomiej Kwieciszewski – top), 121 (pixies), 122 (Steffen Foerester), 173 (Asther Lau Choon Siew – top)

**Photos from www.istockphoto.com (copyright © holder in parentheses):** 1 (Stefan Klein – bottom), 3 (Stephanie Phillips – bottom), 6 (none listed), 20 (Matthew Hull), 24 (Brett Atkins), 27 (Frank Hatcher), 28 (Vladimir Pomortsev – top), 20 (Peter Evans – top), 31 (Jami Garrison), 32 (Kwok Chan), 36 (Brett Atkins), 41 (Lloyd Luecke), 42 (Chris Johnson), 44 (Marco Kopp – upper bottom, Nick Jones – lower bottom), 45 (Thomas Hopson – left, Joe Gough – right), 46 (Jonathan Page – top), 47 (Jessica Jones – top middle, Jeff Gynane – top right, Julie Wax – middle left, Hadleigh Thompson – middle middle, 50 (Wayne Johnson), 51 (Steffan Foerster – bottom), 56 (Dan Schmitt), 58 (Charles Babbitt, Peter Heiss), 61 (Paul Topp – bottom), 66 (Dan Schmitt), 67 (Philip Puleo – top), 83 (Kelly Castilla), 89 (Charles Babbit – left top), 93 (Michael Silverman – top), 96 (Sandra vom Stein), 100 (Dan Schmitt – top, Kit – bottom), 104 (Peter Heiss – top), 112 (Ben Slatter – middle), 146 (Janis Dreosti – top), 151 (Paul Rogers – top, Karina Tischlinger – middle, Dan Schmitt – bottom), 166 (Susannah Skelton – top, Chris Pendleton – bottom), 173 (Nikki Jones – bottom), 180 (Daniel Vice – top)

# INDEX